Arthur Horseman Hiorns

A text-book of elementary metallurgy for the use of students

Arthur Horseman Hiorns

A text-book of elementary metallurgy for the use of students

ISBN/EAN: 9783337276171

Printed in Europe, USA, Canada, Australia, Japan

Cover: Foto ©Paul-Georg Meister /pixelio.de

More available books at **www.hansebooks.com**

A TEXT-BOOK

OF

ELEMENTARY METALLURGY

FOR THE USE OF STUDENTS

By ARTHUR H. HIORNS,

PRINCIPAL, SCHOOL OF METALLURGY, BIRMINGHAM AND
MIDLAND INSTITUTE.

*To which is added an Appendix of Examination Questions, embracing
the whole of the Questions set in the three stages of the subject by
the Science and Art Department for the past twenty years.*

London:
MACMILLAN AND CO.,
AND NEW YORK.
1888.

All rights reserved.

PRINTED AT THE UNIVERSITY PRESS, GLASGOW,
BY ROBERT MACLEHOSE.

PREFACE.

THE following pages constitute a purely elementary treatise on Metallurgy adapted to the capacity of beginners, and dealing rather with general principles than detailed processes, many of which are only suitable for more advanced students. It is hoped that by simplifying the subject in this way, some may be induced to take up its study who have been hitherto deterred by the formidable array of matter presented in the larger manuals.

The work is arranged to meet the requirements of pupils preparing for the elementary stage of the Science and Art Department's Examinations in Metallurgy and includes the examination questions set by the Department for the last twenty years in the three stages of the subject.

The lectures of Professor Chandler Roberts-Austen

at the School of Mines, and the works of Percy, Gruner, Kerl, Phillips, Watts, Bauerman and Greenwood, have been largely drawn upon. The author also thankfully acknowledges his indebtedness to Mr. G. C. Marks, C.E., for his kindness in executing the whole of the drawings and to Mr. J. H. Stansbie for reading the proof sheets.

SCHOOL OF METALLURGY,
BIRMINGHAM AND MIDLAND INSTITUTE,
November, 1888.

CONTENTS.

CHAPTER I.

	PAGE
INTRODUCTION, ...	1

Definition—Metallurgical Terms—Ore—Regulus and Speise—Reduction — Calcination — Roasting—Distillation—Sublimation—Liquation—Scorification—Cupellation—Amalgamation—Occlusion—Cementation, . . . 3

Physical Properties of Metals—Table of Melting Points—Pouillet's Table—Tenacity—Toughness—Malleability—Welding—Ductility—Elasticity—Odour and Taste—Conductivity—Specific Gravity—Hardness—Specific Heat—Alloys—Varieties of Fracture, ... 7

Dry Process—Wet Process—Electro-Chemical Process—Crystallization, ... 13

CHAPTER II.

SLAGS AND FLUXES.

Slags—Silicates—Flux—Refractory Materials—Crucibles—Furnaces—Hot Air, ... 15

CHAPTER III.

FUEL.

Definition—Calorific Power—Calorific Intensity—Pyrometers—Wood—Peat—Coal—Lignite—Anthracite—Petroleum—Prepared Fuel—Wood Charcoal—Peat

	PAGE
Charcoal—Modes of making Charcoal—In Piles—In Kilns and Ovens—Coke—Coking in Kilns—Coking in Ovens—Cox's Oven—Appolt's Oven—Coppée's Oven—Simon-Carvè's Oven—Composition of Gases from Coke Ovens—Desulphurization of Coke, ...	27

CHAPTER IV.

Iron.

Ores of Iron—Chemistry of Iron—Iron and Oxygen—Iron and Phosphorus—Iron and Arsenic—Iron and Sulphur—Iron and Silicon—Iron and Carbon—Alloys of Iron—Properties of Iron—Preparation of Iron Ores—Calcination and Roasting—Direct Methods—Calalan Process—Siemens' Process in Rotator—Indirect Extraction—Blast Furnace—Cast Iron Stove—Cowper's Stove—Whitwell's Stove—Waste Gases—Refining Pig Iron—Dry Puddling—Wet Puddling—Treatment of Puddled Balls—Re-heating—Finery Method, ...	44

CHAPTER V.

Steel.

Definition—Effect of Impurities—Cementation Process—Case Hardening—Cast Steel—Indian or Wootz Steel—Open Hearth Steel—Pernot's Furnace—Ponsard's Furnace—Malleable Cast Iron—Steel Casting—Bessemer Process—Basic Process, ...	73

CHAPTER VI.

Silver—Gold—Platinum.

Silver.

Properties of Silver—Silver Compounds—Alloys of Silver—Ores of Silver, ...	85

	PAGE
Extraction of Silver—Liquation—Amalgamation—Patio or Mexican Method—European Method—Extraction of Silver by Lead—Freiberg Method—Wet Methods—Augustin's Method—Ziervogel's Method,	87
Electro-Plating—Chemical Method of making a Solution—Battery Method,	96

Gold.

Properties of Gold—Alloys of Gold—Ores of Gold, . . 98
Extraction of Gold—Amalgamation—Jordan's Process—Electro Amalgamation—Electro-Deposition of Gold—Gilding, ... 100
Separation of Gold from Silver, etc.—Sulphuric Acid Method—Nitric Acid Method—Chlorine Method, . . . 103

Platinum.

Properties of Platinum—Ores—Wet Method of Extraction—Dry Method—Electro-Deposition, ... 105

CHAPTER VII.

COPPER—ZINC.

Copper.

Properties of Copper—Copper and Oxygen—Copper and Sulphur—Copper and Carbon—Copper and Phosphorus—Ores of Copper, ... 108
Extraction of Copper—Welsh Method—Blast Furnace Method—Kernel Roasting—Wet Methods—Electrolytic Method—Copper Plating, ... 110

Zinc.

Properties of Zinc—Compounds of Zinc—Ores of Zinc, . 123
Methods of Extraction—Old English—Belgian—Silesian—Electrolytic Method—Alloys of Zinc, ... 125

CHAPTER VIII.
Lead.

Properties of Lead—Lead and Oxygen—Lead and Sulphur—Alloys of Lead—Ores of Lead—Extraction of Silver from Lead—Cupellation, ... 130

Processes for the Extraction of Lead—Methods by Roasting and Re-action—In Flintshire—In Brittany—In Spain—In Cornwall—In the Ore Hearth—Blast Furnace Method—In Sweden—At Pontgibaud—In the Hartz—Electrolytic Method—Lead Fume—Softening Hard Lead—Reduction of Litharge—Manufacture of Red Lead—White Lead, ... 137

CHAPTER IX.
Tin.

Properties of Tin—Alloys of Tin—Tinning Iron Plates—Ores of Tin—Tin Smelting—In Reverberatory Furnaces—In Blast Furnaces—Deposition of Tin, . . . 147

CHAPTER X.
Nickel—Cobalt.
Nickel.

Properties of Nickel—Nickel Speise—German Silver—Ores of Nickel—Extraction of Nickel—Dry Method—Wet Method—Nickel Plating, ... 152

Cobalt.

Properties of Cobalt—Alloys of Cobalt—Oxide of Cobalt—Smalt—Zaffre—Printer's Blue, ... 156

CHAPTER XI.
Aluminium.

Properties of Aluminium—Ores of Aluminium—Modes of Extraction—Castner Process—Electrolytic Methods—Cowles' Method—Alloys of Aluminium, . . . 158

CHAPTER XII.

Mercury.

Properties of Mercury—Compounds of Mercury—Amalgams—Ores of Mercury—Modes of Extraction—Almaden Process—Hähner's Process—Alberti's Process—Reduction by Lime, Iron, etc.—Purification of Mercury, . 162

CHAPTER XIII.

Antimony—Arsenic—Bismuth.

Antimony.

Properties of Antimony—Compounds of Antimony—Alloys of Antimony—Ores of Antimony—Extraction of Antimony, 167

Arsenic.

Properties of Arsenic — Oxide of Arsenic — Sulphides of Arsenic—Ores of Arsenic—Extraction of Arsenic—Alloys of Arsenic, ... 170

Bismuth.

Properties of Bismuth—Ores of Bismuth—Alloys of Bismuth, 171

APPENDIX, containing examination questions, . . . 173

LIST OF ILLUSTRATIONS.

FIGS.
- 1-5. Crucibles.
- 6. Dome furnace.
- 7. Boetius' ,,
- 8. Siemens' ,,
- 9. Siemens' gas producer.
- 10. Charcoal pile.
- 11. Coke kiln.
- 12. Beehive coke ovens.
- 13. Appolt's ,,
- 14. Coppée's ,,
- 14a. Simon-Carvè's coke ovens.
- 15. Gjer's calciner.
- 16. Iron scheme.
- 17. Catalan forge.
- 18. The trompe.
- 19. Siemens' rotatory furnace.
- 20. Iron blast furnace.
- 21. Cast iron stove.
- 21a. Cowper's ,,
- 22. Whitwell's ,,
- 23. Refinery.
- 24. Puddling furnace.
- 25. Re-heating ,,
- 26. Ponsard's ,,
- 27. Finery.
- 28. Cementation furnace.
- 29. Pernot's ,,
- 30. Bessemer's converter.
- 31. ,, ,, in vertical section.
- 32. Liquation furnace.
- 33. Freiberg scheme.
- 34. Piltz' furnace.
- 35. Augustin's lixiviation tubs.
- 35a. ,, ,, tub in vertical section.
- 36. Electro-plating vat.
- 37. Hungarian mill for gold ores.
- 38. Oxy-hydrogen furnace.
- 39. Welsh copper scheme.

FIGS.
- 40. Calciner for copper ores.
- 41. Gerstenhofer's calciner, front view.
- 41a. Gerstenhofer's calciner, side view.
- 42. Melting furnace for copper regulus.
- 43. Swedish blast furnace for copper ores.
- 44. Swedish blast furnace for smelting black copper.
- 45. Refining hearth for copper.
- 46. Kiln for kernel roasting.
- 47. Tharsis scheme.
- 48. Old English zinc furnace.
- 49. Belgian ,, ,,
- 50. Silesian ,, ,,
- 51. Arrangement of Pattinson's pots.
- 52. German cupellation furnace
- 53. English ,, ,,
- 54. Plan of the test.
- 55. Flintshire furnace for lead ores.
- 56. Scotch ore hearth.
- 57. Castilian furnace.
- 58. Rachette's furnace.
- 59. Stagg's condenser for lead fume.
- 60. German tin smelting furnace.
- 61. Nickel scheme.
- 62. Aludels for condensing mercury.
- 63. Hähner's furnace.
- 64. Alberti's ,,
- 65. Antimony liquation furnace
- 66. Furnace for subliming oxide of arsenic.
- 67. Bismuth liquation furnace.

ELEMENTARY METALLURGY.

CHAPTER I.

INTRODUCTION.

Metallurgy (from the Greek "metallon," ore, metal; and "ergon," work) is the art of extracting metals from their natural state, and preparing them for various manufacturing purposes. There is a tendency at the present time to expand the subject so as to embrace the various methods in which metals are employed for industrial uses.

Most metals in their natural state differ widely from their condition after subjection to the various smelting and refining operations by which they are separated from other matters associated with them in the earth. Ores of metals occur in two ways, viz., in the metallic state, the metal being free or simply mixed with other metals and in a state of chemical combination, which is by far the more common.

Metallurgy is essentially a chemical art, so that the metallurgist necessarily requires a knowledge of the laws of chemistry. But the physical properties of a metal are as important as the chemical, and as metals cannot be profitably extracted and worked without mechanical contrivances, sometimes of an extensive and complicated kind, the metallurgist should not be ignorant of physics and of the principles of mechanics.

The present work will be confined entirely to the consideration of a limited number of the metallic elements known as the "useful metals," which are—iron, copper, zinc, lead, tin, antimony, arsenic, bismuth, nickel, cobalt, mercury, silver, gold, platinum and aluminium. These metals are chiefly found in nature in the form of compounds which must be decomposed in order to isolate the contained metal. The reduction is generally effected by the agency of heat, so that some knowledge of the nature and properties of fuel is requisite. Most metallurgical operations are conducted at high temperatures, thus requiring furnaces built or lined with some material capable of withstanding excessive heat without melting or decomposing, and the present work would be incomplete without some explanation of the refractory materials employed in making bricks, crucibles, and furnace linings. Further, it is not only necessary to reduce a compound to the metallic state, but also to form and fashion the metal by means of hammers, rolls, presses, etc., in order to prepare it for industrial purposes. The nature of such mechanical contrivances will therefore be briefly alluded to.

A metal may be separated from its ore by--

1. The agency of heat, termed the "dry way."
2. Solution and precipitation, termed the "wet way."
3. The agency of electricity, termed the "electro-chemical way."

Definition of Metallurgical Terms.

Ore.—An ore is a substance containing a metal in its natural state, chiefly as sulphide, oxide, or carbonate, and less frequently as arsenide, chloride, sulphate, phosphate, and silicate. Some metals, such as gold and platinum, are usually found in the free state, termed "native," which is the case occasionally with copper, tin, silver and other metals. Metallic substances rarely occur alone in nature, being generally mixed with the various minerals in which they are embedded, constituting the "vein stuff," "matrix," or "gangue," which may consist of carbonates, such as calc-spar, witherite, limestone; silicates, such as felspar, hornblende and mica; sulphates, such as heavy spar; and fluorides, such as fluor-spar. The gangue is separated by the miner, as far as it is expedient to do so by mechanical operations, such as crushing, sorting, washing, etc., known as "dressing," before the ore is delivered to the metallurgist.

Regulus and Speise.—Certain metals, such as copper and silver, possess a strong affinity for sulphur and may be converted into sulphides by fusion with such bodies as iron pyrites, or the sulphates of barium and lime. The sulphide thus formed is called a "Regulus" or "Matt." In a similar way the metals nickel and cobalt are converted into arsenides by their combination with arsenic; the arsenide is then termed a "Speise." In smelting ores containing iron, copper, nickel, arsenic, sulphur and silica, three products may be obtained, viz., nickel speise, copper regulus and iron slag.

Reduction.—When a metal is isolated from its state

of combination, it is said to be reduced and the agent effecting the reduction is called a reducing agent. Reduction may be partial or complete. Thus, a compound may be reduced from a higher to a lower state of combination, or completely reduced to the metallic state. For the purposes of reduction a metallic compound is generally converted into the state of oxide, if not already in that form. Compounds of metals, like gold, silver, platinum, and mercury, may be reduced by heat alone. Sometimes, as in the case of sulphides, the compound is partially roasted to oxide, and then on raising the temperature the two bodies react on each other with separation of the metal. The chief reducing agents in the dry way are carbon, carbonic oxide, hydrogen and hydrocarbons.

When carbon is employed and the oxide is easily reducible, carbonic acid is formed; but when the temperature is such that carbon will reduce carbonic acid, then carbonic oxide is produced, which is always the case with ores difficult to reduce. Thus :—

$$2PbO + C = CO_2 + Pb_2.$$
$$ZnO + C = CO + Zn.$$

The various processes by which a metal is reduced by the agency of heat is called "smelting."

Calcination.—This is a preliminary operation (resorted to in the case of many ores) by which volatile matter is partially or wholly expelled, by heating the substance to a temperature below its melting point, thus rendering it more porous and more suitable for the subsequent smelting.

Roasting.—When metalliferous matter, especially in a finely-divided state, is oxidized by heating it in contact with air, it is said to be roasted. In some cases, as in the roasting of lead sulphide, the oxidation may be partial, sufficient unaltered sulphide being left to react on the oxide so as to isolate the metal, thus:—

$$PbS + 2PbO = SO_2 + 3Pb.$$

In some cases it is desirable to collect the gaseous product. The chamber containing the metalliferous matter is then often isolated from the chamber along which the flame passes.

When chlorine gas is passed over the substance instead of air, the operation is termed a chlorinizing roasting.

Distillation is the operation of driving off a volatile substance by heat in the form of vapour, and condensing it to a liquid in some cool receptacle. Mercury and zinc are extracted from their ores by a distillation process.

Sublimation is a process analogous to distillation, except that the substance set free is condensed in the solid state, such as arsenic for example.

Liquation.—In some cases one or more of the constituents of a mixture may be separated from the rest by taking advantage of their different melting points: thus, when a mixture of lead and copper is raised to the fusing point of the former, the greater part of the lead flows away or liquates out from the unmelted copper. In this way tin is freed from iron and arsenic; sulphide of antimony and native bismuth from their associated gangues, etc.

Scorification.—The word "scoria" strictly refers to the fusible compound containing one or more useful metals, which is produced by the union of a suitable flux with the extraneous materials of the ore or metalliferous matter, such as silicate of lead ($2PbO,SiO_2$). When such a compound contains only earthy bases, such as lime or alumina, it is termed a "slag," such as silicate of lime ($2CaO,SiO_2$). In this country the term slag is in use for both kinds of such compounds. Scorification is the operation of converting the foreign substances present in a metallic compound into a slag or scoria, and the vessel in which such a change is effected on the small scale is called a scorifier.

Cupellation.—This operation is performed in a vessel made of bone-ash or wood-ash, called a "cupel," and has for its object the removal of base metals by oxidation in conjunction with lead. The vessel being very porous (unlike a scorifier) absorbs the metallic oxides, leaving the unoxidizable metals on the cupel. This method is used in the extraction of gold and silver from ores and other bodies containing them.

Amalgamation is, strictly speaking, the union of mercury with other metals, but the term is applied to the whole process in which mercury is used to extract metals, such as gold and silver, from their ores, the mercury being subsequently driven off by heat, leaving the noble metals in the pure state.

Occlusion.—Graham has shown that a palladium tube when heated is capable of absorbing a large quantity of hydrogen gas, and other experimenters have shown that metals exert a selective influence on gases, thus—

iron absorbs carbonic oxide very readily; silver occludes oxygen, and platinum has a considerable absorptive power for hydrogen and oxygen.

Cementation.—When metallic matter is heated without melting in contact with an oxidizing re-agent, so that certain impurities are oxidized by the gas slowly penetrating the mass little by little, the process is an "oxidizing cementation," as in the production of malleable cast iron.

When wrought iron is strongly heated in contact with carbon or carbonaceous matter, it gradually unites with a portion of the carbon, converting the iron into steel. This is a "carburizing cementation." Cementation then is the reaction which takes place between two bodies without fusion.

Physical Properties of the Metals.

Mercury is the only metal liquid at the ordinary temperature. All the rest are solid and all are opaque, except in very thin slices. Gold leaf, for example, transmits a greenish light. With regard to colour, gold is yellow, copper red, the remaining metals white or greyish white. The high reflective power of metals confers on them a particular brilliancy called metallic lustre. The influence of heat on the different metals is very varied; thus, tin, lead and zinc melt below a red heat; gold, silver and copper require a red heat; iron, manganese, nickel and cobalt an intense white heat; platinum only melts at the temperature of the electric arc or that of the oxy-hydrogen flame.

Table of Melting Points.

Platinum,	—
Iron, Nickel,	about 2000° C.
Steel,	below 2000
Cast-iron,	about 1500
Gold,	,, 1200
Copper,	,, 1100
Silver,	,, 1000
Antimony,	,, 450
Aluminium,	—
Zinc,	,, 420
Lead,	,, 330
Bismuth,	,, 270
Tin,	,, 230
Mercury,	,, −39·4

Some metals, such as zinc, arsenic, and antimony, readily vapourize when heated, arsenic being the most volatile. Arsenic passes directly from the solid to the gaseous form when heated, without being liquefied. If, however, it be heated under great pressure liquefaction takes place. All metals are volatile when raised to a sufficiently high temperature.

The following table by Pouillet will give an idea of the temperatures corresponding to different colours:—

Incipient red heat corresponds to	525° C.	977° F.
Dull red ,,	700	1292
Incipient cherry red ,,	800	1472
Cherry red ,,	900	1652

Clear cherry red corresponds to	1000° C.	1832° F.
Deep orange	,,	1100	2012
Clear orange	,,	1200	2192
White	,,	1300	2372
Bright white	,,	1400	2552
Dazzling white	,,	1500	2732

Metals expand when heated and contract on cooling, and within certain limits the expansion is proportional to the degree of heat. Certain anomalies however exist, thus—Cast iron expands at the moment of becoming solid, and solidified bismuth occupies a larger space than when in the liquid state.

Tenacity.—The property of resisting rupture by traction. The power possessed by different metals of sustaining weights is very diverse, and influences the purposes to which they may be applied. To test the tenacity, wires or rods of equal diameters are taken, and a gradually increasing load applied at one end until rupture occurs. The weight required to produce rupture varies with the molecular condition, and with the manner in which the stress is applied. Cast rods are weaker than drawn ones; fibrous iron has a higher tenacity than crystalline iron; the resistance is greater when the weight is applied at once, than when gradually added. The relative tenacity of metals is as follows:—Lead, tin, zinc, gold, silver, platinum, copper, iron and steel; lead being lowest.

Toughness is the property of resisting fracture by bending or torsion, as shown in the case of the metals copper and lead.

Malleability.—When a body can be flattened under a hammer or between rolls without cracking, it is said to be malleable. If it breaks it is termed brittle. Malleability depends on softness and tenacity. During the working of metals their molecules are forced into unnatural positions and require occasional annealing or softening by heat to bring them to their normal state. Gold is the most malleable of metals, combining the two properties of softness and tenacity in the highest degree. The metals follow in order, thus :—Gold, silver, copper, tin, platinum, lead, nickel, zinc and iron.

Welding.—Many metals can have two clean surfaces joined together by pressure forming one compact mass. This property is possessed by iron at a white heat, but lead and gold will adhere at the ordinary temperature.

Ductility.—When a metal can be drawn into wire, or lengthened by a tensile force, combined with lateral compression, without breaking, it is said to be ductile. The ease with which it may be reduced will depend on its softness, but the thinness will depend on its tenacity, which property has more influence on the ductility of metals than on their malleability.

Elasticity or temporary elongation is an attribute of the harder metals, as well as "sonorousness," which is very marked in some of their alloys.

Odour and Taste.—Many metals when rubbed or raised in temperature emit a characteristic odour, and when applied to the tongue produce a peculiar taste, even when chemically pure. The latter can hardly be due to voltaic action.

Conductivity is the power different bodies possess

PROPERTIES OF METALS.

of transmitting heat and electricity. The conducting power of metals for heat is probably in the same order as for electricity. A very minute trace of an impurity will seriously affect this property. One half per cent. of iron in copper will reduce its conductivity 60 per cent., and a much smaller quantity will make it unfit for many electrical purposes. The conducting power of metals is diminished by a rise in temperature. The following is the probable order of conductivity :—Silver, copper, gold, tin, iron, nickel, lead, platinum, antimony, and bismuth.

Specific gravity is the number which expresses the weight of a body compared with that of an equal bulk of water. This number determines the use of a metal for many purposes—thus, gold on account of its high specific gravity is convenient as a circulating medium, and aluminium on account of its lightness is very useful for small weights.

Table of the Specific Gravity of Metals.

Platinum	21·5	Copper	8·96	Antimony	6·80
Gold	19·5	Nickel	9	Arsenic	4·71
Mercury	13·59	Manganese	8·00	Aluminium	2·67
Lead	11·45	Iron	7·90	Magnesium	1·74
Silver	10·50	Tin	7·29	Sodium	·97
Bismuth	9·90	Zinc	7·10	Potassium	·86

Hardness, like many other physical properties of a metal, is often considerably increased by mixture and by presence of impurities, so that softness in many cases is a test of purity.

Specific heat is the number which expresses the

capacity for heat which a body possesses, compared with that possessed by an equal weight of water. Different bodies varying in this respect will require different amounts of heat to raise equal weights from 0° to 1° C.

Table of Specific Heats of Metals.

Iron	·1138	Tin	·0562
Nickel	·1086	Antimony	·0508
Zinc	·0955	Platinum	·0324
Copper	·0952	Gold	·0324
Silver	·0570	Lead	·0314

Alloys.—By uniting two or more metals in various proportions, an almost infinite variety of modifications may be obtained, possessing more or less the properties of their constituents. The effect of this combination is generally to increase the hardness, lower the melting point and otherwise modify the physical properties of the constituent metals. Thus, a small amount of lead in gold makes it very brittle. Some metals mix when melted and separate on cooling, such as zinc and lead. It is difficult to determine whether there is true chemical combination between the components of an alloy, as the alteration of physical characters, which is the distinctive feature of chemical combination, does not to any great extent take place. In some cases, as when copper and zinc are melted together, great heat is evolved, indicating chemical action, and the stability of alloys is greater in proportion to the chemical dissimilarity of the contained metals.

The action of solvents on alloys is sometimes very different from their action on the constituent metals in the separate state: thus, platinum is insoluble in nitric acid, but is dissolved when alloyed with much silver; silver alloyed with much gold is not affected by nitric acid, but when the silver is in large excess the silver is completely soluble, together with some of the gold.

Varieties of fracture.—The character of the fractured surface of a metal or alloy often affords a general guide as to its properties and adaptability to various uses. Zinc, antimony, spiegel-eisen, etc., are crystalline; grey pig iron, steel, bronze, etc., are granular; wrought iron and nickel are fibrous; copper is finely fibrous and silky. Some malleable metals, such as tin, when heated to a certain point and struck with a hammer, or allowed to fall from a sufficient height, when on the point of melting, assume a columnar structure, and certain brittle alloys break like glass with a conchoidal fracture.

Dry process.—By this term is understood the extraction of metals by the agency of heat, three substances being generally employed, viz., fuel to produce heat, or to produce the metal, and flux to combine with the foreign matter present in the ore.

Wet process.—In this method the metalliferous matter is first dissolved by a suitable solvent, such as the common acids, common salt, ammonia compounds, acid sulphates and certain mineral waters, and then precipitated by a certain re-agent. In some cases the substance is first treated by a gas such as chlorine or sulphurous acid and then by a wet method. Some metals after being brought into a state of solution are precipitated by other metals more positive than themselves; thus, copper is precipitated by iron, and silver by copper.

Electro-chemical process.—In this method the substance is first brought into the liquid state by fusion or by solution, and then decomposed by the passage of a current of electricity. Probably the greatest success has been achieved with copper compounds, because copper when deposited is generally very free from impurities. With metals of less intrinsic value, the power required is too costly, unless cheaper methods of producing electric currents than are known at present can be devised. Even then the deposited metals would probably not be pure like copper.

Crystallization.—The brittle metals such as antimony, bismuth and zinc exhibit a well defined crystalline structure on their fractured surfaces, and malleable metals may have the same structure induced under certain circumstances. All metals appear to crystallize either in the cubic or rhombohedral system.

Metals may crystallize (1) on solidification after fusion, (2) by condensation from the state of vapour, (3) by electrolytic decomposition of metallic solutions. Slow cooling after fusion appears to be the condition most favourable for the formation of good crystals. In some instances a crystalline structure is induced by vibration, as in the case of iron and brass. Some combinations containing different elements may have one constituent separated from the rest by crystallization, such as graphite from cast iron, lead from argentiferous silver, zinc from lead, etc. Many metals which are deposited from their solutions by electricity, such as silver and copper for example, may be obtained in a crystalline condition by using a weak current.

CHAPTER II.

Slags and Fluxes.

Slags, with a few exceptions, are formed by the union of metallic oxides with silica and are termed silicates. They may be stone-like, glassy, or crystalline. Rapid cooling tends to produce the glassy variety, slow cooling, to produce the crystalline condition. The crystalline or stony condition of slags depends on the nature of the constituents. They are generally vesicular from the escape of enclosed gases. Some slags when melted are as liquid as water, others are viscous. If a slag is made too thin, it is liable to contain too much metal; if too thick, the metal does not properly subside. When a slag is free or nearly free from metal it is said to be "clean." With regard to fusibility, etc., Berthier states:—

1. Weight for weight soda fluxes better than potash.
2. A mixture of soda and potash is better than either alone.
3. Alkaline silicates are always glassy.
4. Alkaline silicates make other silicates fusible when mixed with them.
5. Only those silicates of baryta fuse well which contain less baryta than BaO, SiO_2, and less silica than $BaO, 6SiO_2$.

6. With respect to silicates of alumina, $Al_2O_3, 3SiO_2$ and $2Al_2O_3, 9SiO_2$ soften most with heat and any addition of silica or alumina diminishes the fusibility.

7. Alkalies, alkaline earths and the earths, flux silica in proportion to their chemical energy, that is, in the following order :—Potash, soda, lime, magnesia, baryta, strontia, oxide of iron and alumina. Among the common bases alumina is the least fluxing.

A slag consists of two portions, the acid such as silica and the base such as alumina. But alumina may act the part of base in one compound as $(2Al_2O, 3SiO_2)$, and the part of acid in another as in the natural mineral "spinel," ZnO, Al_2O_3. The former would be called a silicate and the latter an aluminate. The aluminate of lime or of magnesia is infusible, but the fusibility is increased by forming double aluminates. Silicates may be classified as follows :—

Monosilicates,	- -	$2MO, SiO_2$ or MS.
Sesquisilicates,	- -	$4MO, 3SiO_2$ or M_2S_3.
Bisilicates,	- -	MO, SiO_2 or MS_2.
Trisilicates,	- -	$2MO, 3SiO_2$ or MS_3.

A flux is a substance added to metalliferous substances to unite with the foreign matter and form a fusible slag. The flux employed in any case varies with the nature of the gangue, which is generally either siliceous or basic. A siliceous gangue requires a basic flux and likewise a basic gangue, a siliceous flux. The gangue of an ore may be self fluxing and require no additional flux. A single silicate with one base is less fusible than a double or multiple silicate with two or more bases. Thus ores of iron containing silicate of alumina, have lime added as a flux, so as to form a double silicate of lime and alumina.

Fluor-spar as a flux acts on silicates in two ways:— 1. By combining with the silicate to form a fusible compound. 2. By decomposing the silicate, forming gaseous silicon fluoride (SiF_4). Fluor-spar is a useful flux for the sulphates of barium, calcium, and strontium, and for bone-ash, but it has no action on sulphides.

Sulphides are for the most part fusible *per se*, but they are generally roasted to oxides. Certain sulphides as ZnS, are made fusible by the addition of iron pyrites. By means of sulphur or arsenic small quantities of copper and nickel may be removed from other oxides in the form of regulus or speise.

Refractory Materials.

In any hearth or furnace where a high temperature is desired, it is necessary that it should be lined inside with a material capable of withstanding the heat and scorifying action of the matter operated upon, without decomposing. Such materials are either used in the natural state, such as silica, alumina, lime, magnesia, oxides of iron and fire-clay, or they are made to undergo a certain preparation. In some cases the natural materials are moulded to the internal shape of the furnace. If they are not of a plastic nature, such as lime and magnesia, then tar or some other binding material is added to give the necessary plasticity. Fire-bricks are generally made of fire-clay mixed with burnt clay and white sand which prevents the bricks cracking, without increasing the fusibility. The composition differs with the purposes for which they are designed. In some cases they are required to withstand a high and prolonged temperature without softening; in others to withstand great pressure; in others to resist the corrosive action of metallic oxides, and in others to withstand great and sudden changes of temperature.

Crucibles, etc., are made of various mixtures of clay in the raw and burnt state, or the same mixed with coke dust or plumbago. A good crucible should be tough, infusible, capable of withstanding sudden changes of temperature without cracking, and should not be readily corroded by metallic oxides. The most infusible crucibles are those made with clays containing the largest amount of silica, and the smallest quantity of lime and oxide of iron. A good crucible may be made with two-thirds fire-clay and one-third burnt fire-clay and coke dust. The refractory nature of fire-clays used for crucibles, fire-bricks, etc., may be tested by moulding a piece in form of a pyramid with sharp edges, drying, and then exposing to a high temperature in a crucible lined with charcoal. If only the edges show signs of fusion, the clay is moderately refractory, if the clay fuses it is useless for high temperatures.

Graphite, blacklead or plumbago crucibles are made of fire-clay mixed with varying proportions of plumbago or coke dust. Instead of using blacklead crucibles, clay ones lined with charcoal paste are often employed. Such a crucible is said to be brasqued (Fig. 1). The paste may be made by mixing treacle with the charcoal, pressing in the mixture and cutting out the central part

CRUCIBLES.

with a knife, or a cavity may be made by a wooden plug.

The following crucibles are in general use for assaying. 1. *French.*—These are of excellent quality, smooth and carefully made but somewhat brittle (Fig. 1 *a*). 2. *London.*— These crucibles have a reddish brown colour, are close in grain, refractory and well resist the corrosive action of metallic oxides (Fig. 2). 3. *Cornish.*—These crucibles are refractory but of a more acid character than the preceding and thus more readily attacked by metallic oxides (Fig. 3). 4. *Hessian.*—These are useful crucibles, refractory, not readily corroded, but somewhat frangible and liable to crack with sudden changes of temperature (Fig. 4). In making large crucibles for melting brass, steel, etc. (Fig. 5), a mixture of Stourbridge clay with other refractory clays, such as China clay, Derbyshire clay, and plumbago, coke dust and burnt clay are employed. Good plumbago crucibles resist sudden changes of temperature without cracking and may be used several times until they are worn too thin to bear the weight of the metal. Crucibles of all kinds should be carefully annealed before using, by heating in an inverted position over the furnace.

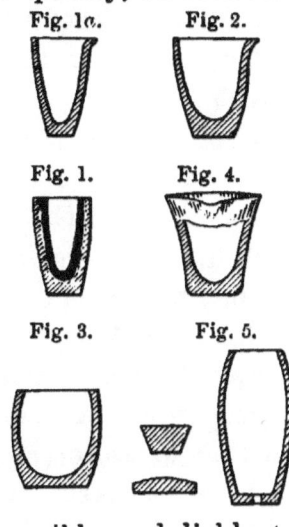

Refractory materials are either acid, such as ganister and Dinas clay; neutral, such as fire-clay; or basic, such as dolomite, bauxite, alumina, etc.

A substance is called acid, neutral, or basic, when the acid is greater, equal to, or less in equivalence than the base. This distinction will be seen by reference to the composition of the above bodies :—

Ganister = SiO_2 89·5, Al_2O_3 4·8, FeO_2 ·8, CaO ·1, K_2O ·1.

Dinas Clay = SiO_2 98·3, Al_2O_3 ·7, FeO ·2, CaO ·2, K_2O ·1, OH_2 ·5.

Kaolin, which is the purest form of fire-clay, contains 40 per cent. SiO_2, 45 Al_2O_3, and 5 OH_2.

Dolomite ($CaO,CO_2 + MgO,CO_2$). In this case the carbonic acid is removed by heat, leaving the oxides of calcium and magnesium, which are entirely basic.

Bauxite = Al_2O_3 52, Fe_2O_3 27·6, water 20·4.

Fire-clays are essentially hydrated silicates of alumina, which resist exposure to high temperatures without melting or softening. They contain varying amounts of lime, magnesia, oxide of iron, potash, etc., and some mechanically mixed silica. The tendency to fuse or soften increases with the amount of oxide of iron and potash. The plastic property, or the power of being moulded into various shapes, which clays possess, is due to the chemically combined water. The mineral Kaolin, mentioned above, may be considered as pure clay and probably forms the basis of all clays.

Furnaces.

A hearth, forge, or furnace is an arrangement for conducting metallurgical operations by the agency of heat. The inside is generally lined with refractory material and the outside constructed of ordinary building material. The bed of a furnace as a rule is lined with basic material, which does not readily act on metallic oxides; but the roof, which only comes in contact with the flame, is made of refractory acid materials, such as Dinas clay.

All furnaces may be classified under three types:—
1. Without an independent fireplace as an iron blast furnace. 2. With an independent fireplace as a reverberatory furnace. 3. Closed muffles as in zinc distilling. They may be used with a natural draught, the air being aspirated by means of a chimney; or the air may be forced in with bellows or similar means. In both cases, the heat developed depends on the calorific power and weight of fuel burned in a given time; also on the mode of combustion, for if the carbon is completely burned to carbonic acid, more than three times the heat will be produced than when only carbonic oxide is formed.

We will call No. 1 the barrel type; the ore, fuel and flux being placed together in the same receptacle. No. 2, the reverberatory type, contains two compartments, one in which the fuel is burnt, and the other where the metal or mineral matter is operated upon, called the "laboratory." The barrel type is favourable for reducing-fusion, but the reverberatory type is more oxidizing and suitable for such operations as require the combined action of heat and air. The fuel may be solid or gaseous, in which latter case the arrangement is called a "gas-furnace." The chamber where the gas is produced is called a gas generator or producer, and by this agency temperatures are now obtained previously unknown.

Besides the fireplace and laboratory there is another

essential structure connected with a reverberatory furnace, called the stack or chimney. It may be used either for creating and maintaining the draught, or for carrying off the products of combustion. The chimney is generally connected with the laboratory by a "flue" of varying length and more or less horizontal. The draught is regulated by a damper either in the flue or on the top of the chimney. Some furnaces have a fireplace in the centre and the flame passes right and left, as in the cementation of iron to form steel. (See Fig. 28.) These are called "gallery" furnaces. Some have the fireplace more completely isolated than in the ordinary reverberatory. The laboratory space is covered with a dome which is pierced with one or more openings connected with the chimney. The fireplace is under or at the sides of the laboratory, and the gaseous products of combustion penetrate into the laboratory by a number of openings (Fig. 6).

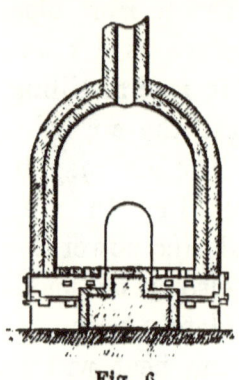

Fig. 6.

The atmosphere of a reverberatory furnace will depend to some extent on the depth of fuel on the grate; for when this is too great the combustion will be incomplete and carbonic oxide with free hydrogen will be largely formed, producing a more or less reducing atmosphere. With a thin layer of fuel and good draught, the atmosphere will be oxidizing, carbonic acid and water being chiefly produced.

Hot Air.

If the air supplied to a furnace be previously heated, the temperature of combustion will be increased, and although fuel is used in heating the blast there is still a margin on the side of economy, for more complete combustion occurs and a more rapid working is realized. If the waste heat be used in heating the air, then the saving is still greater. Two principles are applied for heating air, viz., "Conduction" and "Regeneration."

The principle of conduction is employed when the air is passed through cast-iron pipes which are heated externally. (See Fig. 21 *a*.) Or the walls of a gas producer may be made hollow so that the air for combustion of the gas may pass through such passages, as in the Boetius method (Fig. 7). Or the air may be heated by

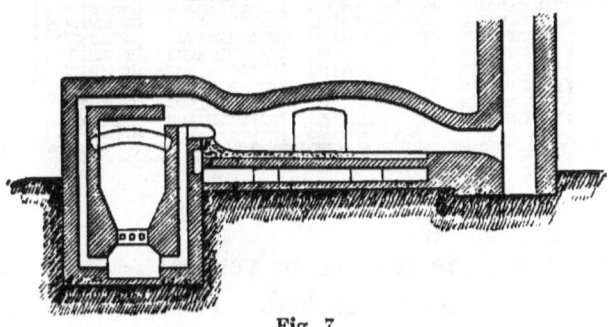

Fig. 7.

passing through a channel underneath the bed of the laboratory as in the Bicheroux furnace. Or what is termed a "recuperator" may be employed. In this arrangement the waste gases from the furnace, and air, pass in opposite directions through alternate channels in a specially made brickwork chamber. (See Fig. 26.)

The system of heating on the regenerative principle was developed by Sir William Siemens and so named by him because the waste heat was restored again to the furnace. The regenerators are chambers of open refractory brickwork built in pairs, two pairs being required for each furnace, each pair being used alternately for absorbing the heat of the gaseous products from the furnace, and heating the gas and air required for combustion.

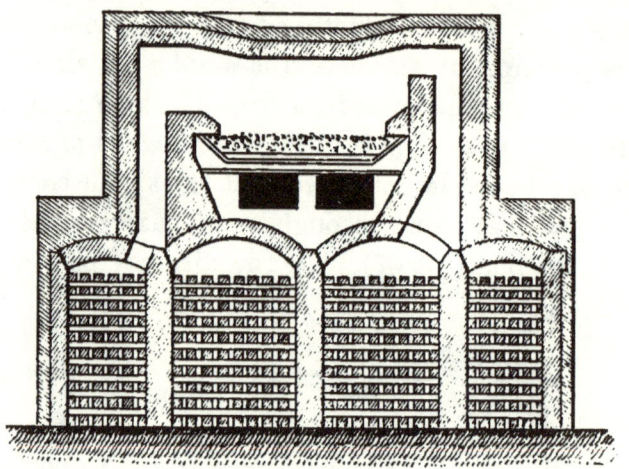

Fig. 8.

Fig. 8 shows the furnace in vertical section, with the regenerators beneath the bed. The larger chamber in each pair is used for heating the air and the smaller for heating the gas. By means of a reversing valve the waste gases pass to the right or left pair at will. When the waste gases are passing down through the right pair, the cold air and gas are passing up through the left pair, the direction being reversed when sufficient heat has been absorbed.

The bed of the furnace is formed of sand, consolidated by pressure and strong heating, and supported by a framework of cast-iron. The roof slopes from each end to the centre, thus giving a more plunging flame.

The Gas Producer is a nearly rectangular chamber lined with fire-brick (Fig. 9). The side A is formed of iron plates lined with fire-bricks having a step grate B and wrought iron bars C. The fuel is charged through the hopper D. The gas passes up the pipe E, which is cased with iron, and into a horizontal wrought iron pipe which conveys it to the regenerator. The combustible portion of the gas consists chiefly of carbonic oxide (CO) called "air" gas. When a jet of steam is introduced into the generator the gas liberated contains hydrogen; it is then called "water" gas. The gases generated in gas producers for metallurgical purposes contain from 25 to 34 per cent. carbonic oxide, 55 to 60 per cent. nitrogen, the remaining portion consisting of carbonic acid, hydrogen, and hydrocarbons.

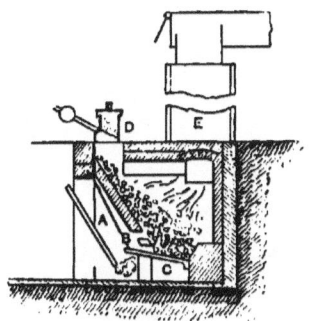

Fig. 9.

The effect of water vapour or steam on incandescent carbon is to form hydrogen, marsh gas, carbonic oxide, and carbonic acid. The effect of oxygen is to produce a mixture of carbonic oxide and carbonic acid, the relative proportion of which will depend on the amount of oxygen admitted. Now hydrogen is a reducing agent, and will act on carbonic acid, reducing it to carbonic oxide; so that a certain proportion of steam admitted with the air in generating producer gas tends to increase the amount of carbonic oxide, and thus enhance the value of the gas as a combustible. The amount of steam admitted should never exceed 10 per cent. of the air, otherwise the cooling effect of the water vapour will be greater than the heating advantage gained by the liberated hydrogen.

CHAPTER III.

Fuel.

By this term is meant—substances that may be burned in air to give heat capable of being applied to economic purposes. The two chief elements employed are carbon and hydrogen. The latter in fuels being always associated with carbon. In all fuels containing carbon, hydrogen, and oxygen, the proportion of hydrogen may be equal to, or greater than, but never less than that required to form water with the oxygen. The amount of hydrogen united with oxygen is not available as a source of heat, so that the excess of hydrogen over this amount is called "disposable" hydrogen, and the amount in combination, the "non-disposable" hydrogen.

Calorific power.—The amount of heat a body is capable of yielding when completely burned is termed its calorific power. The calorific power of different bodies is given in the following table:—

Wood charcoal,	8080
Gas carbon,	8047·3
Artificial graphite,	7762·3
Native graphite,	7796·6
Diamond,	7770·1
Carbonic oxide,	2403

Carbon when burnt so as to form carbonic oxide,	2473
Marsh gas (CH$_4$),	13063
Olefiant gas (C$_2$H$_4$),	11857
Native sulphur,	2249
Amorphous silicon,	7830
Crystallized silicon,	7540
Phosphorus,	5747
Iron,	4134
Hydrogen,	34462

The number 8080 for wood charcoal signifies that 1 lb. of that body when completely oxidized to carbonic acid will produce sufficient heat to raise 8080 lbs. of water 1° C.; and so on for the rest. The calorific power of a fuel containing carbon, hydrogen, and oxygen would be the sum of the calorific powers of the carbon and that of the disposable hydrogen.

The calorific power of a compound of carbon and hydrogen, such as C$_2$H$_4$, may be calculated thus:—

$C_2 = 24$ ⎫ The atomic weight of carbon being 12 and
$H_4 = 4$ ⎭ that of hydrogen 1.
____
 28

Then if 1 lb. be taken, $\frac{6}{7}$ will be carbon and $\frac{1}{7}$ will be hydrogen.

$\frac{6}{7}$ of 8080 = 6925·7
$\frac{1}{7}$ of 34462 = 4923·1
_____
11848·8

Many hydrocarbons yield a less calorific power by experiment than by calculation, probably owing to the heat absorbed in their decomposition. Experiment and calculation perfectly agree, only when the two elements are mechanically mixed, and not when they are chemically combined.

Calorific intensity.—The pyrometric degree of heat obtained when a body is completely burnt is termed its "calorific intensity." This is the actual temperature of the products of its combustion, and will vary with their weight, specific heat, pressure, and the atmosphere in which the combustion takes place. In the case of hydrogen, the product is a condensable vapour, and it is necessary to deduct from its calorific power the amount of heat rendered latent by the water vapour produced.

Suppose 12 grains of carbon to unite with 32 grains of oxygen to form 44 grains of carbonic acid, or 1 grain of carbon to unite with 2·6 grains of oxygen to form 3·6 grains of carbonic acid, thus—

$$C + O_2 = CO_2,$$

and the whole heat produced absorbed by this CO_2; that the pressure is constant and that the specific heat of CO_2 = ·2164; then the calorific intensity = $\dfrac{8080 \times 1}{3\cdot 6 \times \cdot 2164}$ = 10174.

Now suppose the carbon burnt in ordinary air instead of oxygen. In this case nitrogen has to be heated without yielding anything to the calorific power. One part of oxygen in air is mixed with 3·35 parts of nitrogen, so that the weight of nitrogen to be heated when 1 part of carbon is converted to carbonic acid is 2·6 × 3·35 = 8·93. The specific heat of nitrogen is ·244.

Calorific intensity = $\dfrac{8080 \times 1}{(3\cdot 6 \times \cdot 2164) + 8\cdot 93 \times \cdot 244}$ = 2718° C.

The calorific intensity of carbonic oxide (CO) when burned in oxygen according to the equation,
$$CO + O = CO_2,$$
is 7073. One part of CO forms 1·57 parts of CO_2, then the calorific intensity $= \dfrac{2403}{1·57 \times ·2164} = 7073$,

and in air $= \dfrac{2403}{(1·57 \times ·2164) + (1·91 \times ·244)} = 2984$.

In furnaces, heat is lost by radiation, conduction, and dissociation, as well as by imperfect combustion.

Pyrometer.—The heat of a furnace is measured by an instrument called a "pyrometer." It must be capable of giving a constant indication for the same temperature and must not change with use.

1. Wedgwood's pyrometer is based on the shrinkage of dehydrated clay. Spherical pieces with one side flat and of equal size are employed. He assumed that the shrinkage was proportional to the temperature.

2. Daniell's pyrometer is based on the principle that the expansion of a platinum or iron rod is proportional to the temperature.

3. Siemens' pyrometer depends on the increase of electrical resistance in a conductor, such as platinum, when heated.

4. Byström heats a ball of platinum in the furnace whose temperature is to be measured, and then causes it to roll down an inclined clay tube into water, and computes the temperature from the rise in temperature of the water.

Wood.

Wood is composed of organic tissue called cellulose, and a little other organic matter. Cellulose may be represented by the formula $C_6H_{10}O_5$, and contains 44·44 per cent. carbon, 6·17 per cent. hydrogen, and 49·39 per cent. oxygen. The inorganic matter which forms the ash when wood is burnt varies from 1 to 2 per cent., and consists chiefly of carbonates of potash and lime. When wood is well air dried it retains 18 to 20 per cent. of water, which may be removed by heating it at a temperature of 120° C. Wood for fuel should be of mature growth, and felled when most free from sap.

Peat.

Peat is the product of the alteration of various vegetable substances, chiefly mosses, under the combined influence of air and water, which induces a slow decomposition. The deeper the peat, the darker and denser it is, and the less the remains of vegetable structure. Peat will absorb and retain as much water as wood. It contains more carbon and less oxygen and hydrogen than wood, but has contracted several impurities from the earth, the ash reaching from 5 to 6 per cent. The ash consists of carbonates, silicates, sulphates, and phosphates of potash, soda, lime, alumina, magnesia, and oxide of iron. Peat is sometimes prepared for metallurgical purposes by being strongly compressed.

Coal.

Coal is of vegetable origin, and resembles wood and

peat in character, the more modern the formation in which it is found. It may be divided into three distinct classes—Lignite, Coal, and Anthracite—the first being the newest formation and the last the most ancient. As the series descends, the oxygen diminishes and the carbon increases; but the chief characteristic of different kinds is the ratio of hydrogen to oxygen, and the carbon left on distillation in a closed vessel. In lignites the ratio $\frac{O}{H}$ varies from 6 to 5, and the coke is below 50 per cent. In coal the ratio $\frac{O}{H}$ varies from 4 to 1, and the coke from 50 to 90. In anthracite the ratio $\frac{O}{H}$ is 1 or less than 1, and the coke exceeds 90 per cent.

Lignite.—Four types may be distinguished—fossil wood, earthy lignite, dry lignite, and bituminous lignite. Fossil wood is that variety where the woody structure is still apparent, while bituminous lignite more or less resembles ordinary coal, into which it insensibly passes.

Coal is distinguished from lignite by a more pronounced black colour, greater density, more friability, more lamellar structure, absence of woody fibre, a greater yield of coke, less water in the products from distillation, and that rather basic than acid, and the absence of the disagreeable odour of lignite when burning. In the natural state, coal contains little water and is less hygroscopic in air than wood or lignite.

Coal, by its appearance, properties, and composition, forms a continuous series from lignite to anthracite, with no definite line of separation. The proportionate yield of coke is from 50 to 90 per cent., and the elementary composition as follows :—

Carbon, - - -	75 to	93
Hydrogen, - - -	6 to	4
Oxygen, - - -	19 to	3
	100	100

The first column belongs to coal bordering on lignite, the second column to that approaching anthracite. Between these extremes are varieties having intermediate properties.

Coal may be conveniently divided into five classes based on the yield of coke and action in the fire :— 1. Long flaming coal; 2. Gas coal; 3. Forge coal; 4. Caking coal; 5. Lean or anthracite coal.

Anthracite coal is generally friable, of a pure black colour and lustrous, which varies with the hydrogen present. Much ash makes it harder, denser, and less brilliant in appearance. The combustibility and volume of flame depend on the amount of volatile matter present. The earthy matter or ash is chiefly clay, more or less siliceous and ferruginous, with varying amounts of carbonate and sulphate of lime, a little alkali, and occasionally some phosphates and arseniates. The colour of the ash varies from white to red according to the amount of oxide of iron, the coloured ash being the most fusible. This forms pasty masses when the coal is burnt, called clinkers, which choke up the bars of the fireplace. The calorific power increases from flaming coal to anthracite coal. Sulphur is always present in coal, either as sulphate of lime or iron pyrites, or both, and in combination with the organic constituents of the coal.

Anthracite is compact, of a deep black colour, lustrous, bordering on the metallic, brittle, breaks with an uneven or conchoidal fracture, has a high degree of hardness and cohesion; it inflames and burns with difficulty, giving an almost smokeless flame. Some varieties easily decrepitate. It yields neither water nor bitumen in sensible quantities when heated, and the fragments neither fuse, change their lustre, nor cake together.

Petroleum.—In various parts of the earth are found combustible bodies similar to "petroleum." They are solid, liquid, and gaseous hydrocarbons, rich in carbon. Marsh gas and olefiant gas are gaseous. Petroleum proper is liquid. The asphalte of Mexico is a kind of solid petroleum which fuses below 100° C. An analogous compound, "Boghead" of Scotland, which is a bituminous schist, is richer in bitumen than ordinary coal. When the earthy matter diminishes to the proportion present in ordinary coal, it takes the name of "cannel" coal. The natural mineral called jet occupies an intermediate place between coal and petroleum. It is a bituminous lignite, and disengages, when heated, 55 to 60 per cent. of volatile matter, containing 5 to 6 per cent. of hydrogen.

Prepared Fuel.

Certain fuels, such as soft peat, coal dust, etc., are unsuitable for use in the natural state, but may be prepared by compression, as in peat; or caked, as in coal dust, by mixing with coal tar, pitch, or other cementing material. Wood and peat have their calorific powers increased by artificial drying.

Wood charcoal.—When wood is heated to a temperature of 400° C. out of contact with air; water, acetic acid, tar, carbonic acid, carbonic oxide, hydrogen, marsh gas, etc., are given off, and a black, sonorous, hard mass of charcoal is left. At a lower temperature it is more or less brown, feebly sonorous, but more tenacious than black charcoal. Charcoal always retains some oxygen and hydrogen, but the amount is less the higher the temperature employed in its production. The ash left on combustion is 3 to 4 per cent., and of the same character as that of wood.

Peat charcoal.—The shape of the peat is preserved like wood, when heated in closed vessels. It is black in colour, porous, soft when prepared from peat in its native state, and more compact and dense when prepared from compressed peat. It retains some volatile elements, like wood charcoal.

When steam is passed over red-hot charcoal, hydrogen, carbonic oxide, carbonic acid, and a little marsh gas are formed.

Modes of Making Charcoal.

1. In piles.—The ground should be dry, well sheltered, and near a water supply. The bed is slightly inclined from the circumference to the centre. One or three stakes are first driven in the ground in the centre of the circle, long enough to reach above the upper extremity of the intended mound. Around this, the wood cut into suitable lengths is packed as closely as possible, being placed in a vertical or horizontal direction, and all irregular spaces filled up with small twigs. The wood is then encircled and covered with branches, and the upper portion with turf and charcoal dust, leaving a small part open at the base for the escape of aqueous vapour during the first stage. The space between the three stakes is filled with readily-inflammable wood, which is ignited to start the operation. The wood in the central part of the pile is charred first, then covered up, the combustion proceeding from top to bottom and from the centre to the outside of the heap. Vents are made at suitable intervals, commencing at first near the top, and closed when the carbonization in that region is complete, which is judged by the pale blue colour of the smoke (Fig. 10).

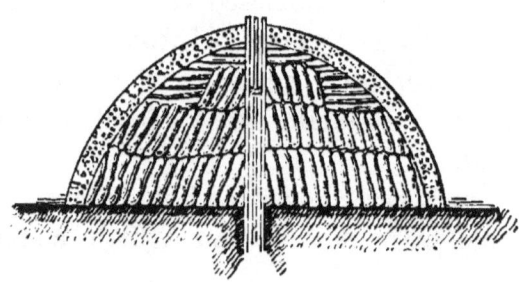

Fig. 10.

2. In some cases rectangular piles are used; in others the charring is effected in kilns and in ovens. The heat by which the carbonization of wood is effected in piles is produced by the combustion of charcoal itself, and not by the burning of the products of carbonization. The yield varies with the nature of the wood, its age, dryness, and the temperature and mode of charring. By

volume the yield varies from 50 to 75, and by weight from 15 to 28 per cent. Peat is carbonized in much the same way as wood—in piles, kilns, and in ovens.

Coke.

Coke is obtained when coal is strongly heated in a closed space or vessel with a limited supply of air. It may be porous and light, or dense and compact; it is sometimes soft and tender; at others, hard and resisting. The colour varies from black to grey, the lustre being in some cases dull, in others, bright and almost metallic. The quality depends on the coal and mode of coking. Nearly all coals contain iron pyrites (FeS_2), and a part of the sulphur remains in the coke. The coke from caking coal has the appearance of a porous, partially melted mass, which is grey in colour, with a semi-metallic lustre. The coke from flaming coal is lighter and more friable. The rapid carbonizing of a small charge gives a lighter and more friable coke than the slow baking of a heavy charge. Dry coke with little ash is lighter than water. It is hygroscopic like charcoal, but in a less degree. Coke is less inflammable and less combustible than charcoal, but produces a higher temperature on burning.

Coking in piles.—Coking is effected in circular and rectangular piles, like charcoal. A rough chimney of loose bricks is first constructed, 6 feet high, capped with a damper, and the coal arranged around it, making a mound 30 feet in diameter at the base. The whole is then covered with wet slack, except a space around the bottom. Ignition is effected by placing live coals on one side of chimney, near the top, and continued by opening vents in different parts, as in charcoal burning. The coking is completed in about six days.

Coking in kilns.—Two parallel walls (AA, Fig. 11), 5 feet high and 8 feet apart, lined with fire-brick, are first built with a series of openings (BB), 2 feet apart, and the same distance from the floor, so arranged that those on one side are opposite to those in the other wall; and from each of these ascends a vertical flue (CC). Any of these flues may be stopped by closing with a tile so as to divert the current in any given direction. To charge the kiln, one end is bricked up, wet slack is wheeled in, spread in layers and stamped down, reaching nearly to the level of BB. Wood stakes are now placed across, the ends reaching into the corresponding opening in the opposite wall. The coal is then wheeled in, watered, stamped down, the other end bricked up, and the whole covered with loam. The stakes are now withdrawn, and the kiln lighted by inserting inflammable sticks. The charring is completed in about eight days. The whole is allowed two days to cool, and the coke is then withdrawn.

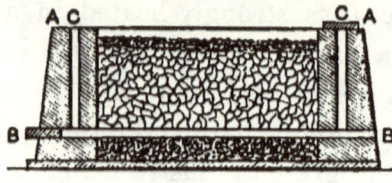

Fig. 11.

Coking in Ovens.

The beehive oven (Fig. 12) is a somewhat circular chamber of brickwork with an arched roof, having a chimney

COKE OVENS. 39

opening at the top for the escape of the products of combustion and vapours. The cavity is about 9 feet to 10 feet in diameter, and 3 feet to 4 feet high. The charge is introduced through a doorway in front, about $2\frac{1}{2}$ feet square, through which the charge is also withdrawn. These ovens are gen-

Fig. 12.

erally built in two rows back to back, with a charge of three tons in each, which reaches up to the springing of the dome of the roof. When the charging is complete, the doorway is loosely filled up with bricks, through the openings of which the air can pass. Supposing the oven to be hot from a previous charge, in three hours the lower holes are closed, and in twenty-four hours the upper ones are closed. The oven is now allowed to remain twelve hours with the chimney open. When the flame ceases, the damper is closed, and the oven allowed forty-eight hours to cool. The charge is then withdrawn by means of a large shovel suspended by a crane, and the hot coke quenched with water.

Cox's oven is a nearly rectangular fire-brick chamber of the beehive type, closed with an iron door lined with fire-brick. The coking space has a double arched roof, forming a space through which the gaseous products pass before reaching the chimney, so as to utilize some of the waste heat. At each side of the door is an opening leading by a flue to openings in the back of the chamber, for the introduction of air heated by passing through this flue, thus raising the temperature of the oven. The coking chamber is slightly wider in front, and the floor also slopes from back to front, which facilitates the withdrawal of the charge.

Appolt's oven.—Eighteen vertical retorts of brickwork are built in two rows of nine each, the whole being contained within four brickwork walls. Each retort is surrounded by an air space, 8 to 10 inches wide. The retorts (Fig. 13) are taper, measuring 1 foot 1 inch by 3 feet 8 inches at top, and 1 foot 6 inches by 4 feet at base, and 16 feet high. Each is provided with a cast-iron door at the bottom, opening into an arched vault, into which the coke is dropped at the conclusion by opening this door. The air spaces surrounding each retort communicate with one another, forming one large divided chamber, which communicates with the inside of the retorts by openings in the brickwork. It is in this divided chamber that combustion of the products of decomposition of the coal takes place, air being admitted through holes in the outer sides of the kiln. The oven is charged with coal at the top, and the time of one operation, starting with hot retorts, is about twenty-four hours.

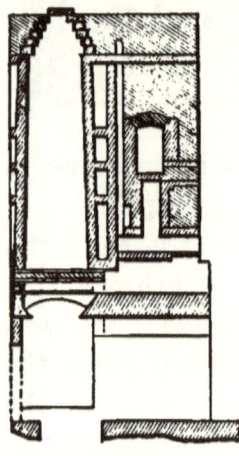

Fig. 13.

Coppée's oven (Fig. 14).—This oven is constructed on the Appolt type, being heated by combustion of the waste gases on the exterior. It is only used for finely divided coal, its chief advantages being rapid coking and an increased yield. Some coals not sufficiently bituminous for coking in an ordinary oven may be coked in the

Coppée oven. There is some saving of labour and a utilization of the waste gases. The gaseous products pass through openings in the upper part of the vaulted roof of each oven into vertical side flues, into which air is admitted. After combustion, the flame passes to a bottom flue, giving up a portion of its heat to the bottom of the oven.

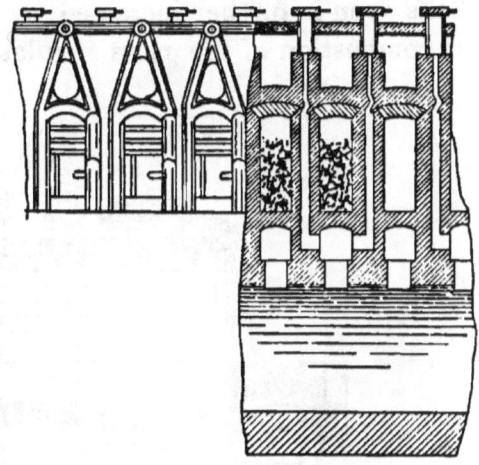

Fig. 14.

Simon-Carvè's oven (Fig. 14a).—Each oven is a long, narrow, and high brickwork chamber, A, resembling that of Coppée, with horizontal flues penetrating the side walls, as shown in dotted outline, BB. These flues communicate with two horizontal flues, CC, running under the floor of each chamber. The products of combustion from the fireplace, E, pass through CC, and then ascend to the uppermost of the side flues, BB. From thence they traverse backwards and forwards along the sides of the chamber, finally passing into the flue D, and thence into the chimney. The products of carbonization are drawn off by an exhauster through F and the valve G into condensers, where the gases are freed from the ammoniacal liquors and tarry oils with which they are associated. The gases are then brought to the fireplace, E, by a pipe, p, and ignite as they pass over the burning fuel. The resulting flame then traverses the bottom and sides of the oven as described before. The temperature of the oven is further increased by admitting hot air for the combustion of the gases circulating round the oven.

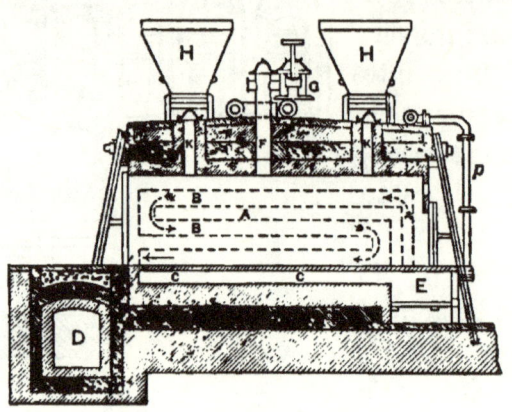

Fig. 14 a.

The coal for coking is charged through the hoppers, HH. When the charging is complete, the passages, KK, are closed, and the doors closing the ends of the ovens carefully luted to prevent admission of air. This prevents loss of coke by combustion at the surface of the charge, and 15 per cent. more coke is obtained than is produced in an ordinary beehive oven.

Composition of gases.—The composition of the gases liberated from coke ovens during coking varies with the time and mode of coking. The mean of some experiments by Ebelmen gave 10·93 per cent. of carbonic acid, 3·42 carbonic oxide, 1·17 marsh gas, 3·68 hydrogen, and 80·80 nitrogen, the first and last being non-combustible. The waste gases of coke ovens have been utilized for heating the ovens themselves, heating boilers, and for the recovery of tar, ammonia, etc.

Coke may be desulphurized: 1. By heating to redness, and passing through it a stream of superheated steam; 2. by adding compressed oxygen; 3. by mixing lime with the coal before coking. The last method appears to be the most successful, and although the lime containing the sulphur remains with the coke, the sulphur has little tendency to pass from the lime into the metal which is being heated in contact with the fuel. The chemical changes in the three methods may be represented thus—

1. $3S + 2OH_2 = SO_2 + 2SH_2$.
2. $S + O_2 = SO_2$.
3. $S + CaO + 3O = CaSO_4$.

CHAPTER IV.

Iron (Fe).

Although iron is very widely distributed throughout the earth, the minerals constituting the workable ores of iron belong to a limited class, consisting chiefly of oxides and carbonates, mixed with varying proportions of limestone, silica, clay, or bituminous matter, and frequently with small quantities of phosphates and sulphides. Sometimes arseniates and titanates are present.

Native iron probably occurs only in meteoric masses.

Magnetite, or magnetic iron ore, is the richest ore of iron. Chemical formula, Fe_3O_4. It occurs crystallized in the cubic system and massive; it is black or iron-grey in colour, brittle, magnetic and sometimes polar, and gives a black streak. When pure it contains 72·41 per cent. of iron.

Franklinite is somewhat similar to magnetite, and contains about 45 per cent. of iron, 10 per cent. of manganese, and 20 per cent. of zinc.

Red hæmatite occurs in a variety of forms. Chemical formula, Fe_2O_3. It contains when pure, 70 per cent. of iron. Its colour varies from deep red to bluish grey, and it gives a red streak. The following are the different varieties:—*Specular iron ore,* of a bluish grey

IRON ORES.

colour, and crystalline. *Micaceous ore* is scaly or foliated, and used for red paint. *Kidney ore* occurs in hard, compact nodular masses. *Red ochre* is a compact earthy variety. *Puddlers' ore* is a compact form used as a fettling for furnaces.

Ilmenite, or titaniferous iron ore, crystallizes similarly to hæmatite, generally occurs massive, of a dead black colour, and gives a brownish streak. The typical chemical formula may be taken as FeO, TiO_2.

Brown hæmatite is a hydrated ferric oxide. Chemical formula, $Fe_2O_3 + xH_2O$. It varies in colour from blackish to yellowish brown, and gives a yellowish brown streak. The following are the different varieties:—

Göthite occurs crystallized, scaly and fibrous, with a colour varying from rust-yellow to black. Chemical formula, Fe_2O_3, H_2O. It contains 63 per cent. iron when pure. *Limonite, or brown iron ore,* is compact and earthy. Chemical formula, $2Fe_2O_3, 3H_2O$. Bog ore is also a brown hæmatite.

Spathic iron ore, Siderite, Clay ironstone, Blackband, and Cleveland ironstone are different varieties of the carbonate of iron ($FeCO_3$). *Siderite* when pure yields 48·27 per cent. of iron. It varies in colour from yellow to brown, and has a pearly lustre. It often contains oxide of manganese, and when smelted produces "spiegel-eisen." It gives a white streak. *Clay ironstone* is the carbonate mixed with clay, and *blackband* is the same compound containing bituminous matter. *Cleveland ironstone* is a less pure variety of clay ironstone, containing a large quantity of phosphates.

Iron pyrites is only used as a source of iron after the

sulphur has been removed in the manufacture of sulphuric acid, when the resulting oxide of iron is known as "Blue Billy."

Chemistry of iron.—Atomic weight, 56. Pure iron is not used in commerce, but it may be obtained by reducing the pure oxide or carbonate with hydrogen, or by electrolysis of a solution of ferrous chloride or sulphate, or by re-melting the best malleable iron with nitre. The red oxide may be reduced by hydrogen at a temperature of 400° C., producing iron in the form of a black powder, while at a high temperature a silvery white sponge is formed. It melts at about 2000° C., is powerfully magnetic, and forms two classes of salts—ferrous and ferric. Its specific gravity is 7·87.

It oxidizes in moist air, forming hydrated ferric oxide ($Fe_2O_3 + Aq$), which, by electrical action on the iron, causes the oxidation gradually to penetrate the whole mass. When heated to redness in air, the black oxide (Fe_3O_4) is formed. Hydrochloric acid attacks iron, forming ferrous chloride ($FeCl_2$). Sulphuric acid dissolves it, forming ferrous sulphate ($FeSO_4$). Fuming nitric acid has no action on iron, but the ordinary acid dissolves it with the formation of ferrous nitrate ($Fe2NO_3$).

Iron and Oxygen unite in three proportions. 1. Ferrous oxide (FeO), which rapidly oxidizes in contact with air, and unites with acids to form ferrous salts. 2. Ferric oxide (Fe_2O_3), which is red in colour and occurs in nature as hæmatite. It unites also with water, forming ferric hydrate. 3. Magnetic or black oxide of iron (Fe_3O_4)—this forms the richest ore of iron, and from which some of the purest iron is obtained.

Iron and Phosphorus unite direct at a red heat, forming phosphide of iron. When oxide of iron is reduced by carbon in presence of an earthy phosphate, phosphorus is separated and unites with the iron. Wrought iron containing $\frac{1}{10}$th per cent. of phosphorus is not much affected in tenacity, but is harder; $\frac{1}{2}$ per cent. makes it cold short, and 1 per cent. makes it very brittle. The effect of phosphorus on iron is to impart a coarsely crystalline structure, diminish its strength, increase its fusibility, and render it cold short. The presence of phosphorus in cast iron diminishes its strength, but on account of its fusibility is useful in making fine castings.

Iron and Arsenic readily unite when heated together. Arsenic makes iron red short, white and very brittle.

Iron and Sulphur combine to form sulphide of iron. The effect of sulphur in small quantities is much the same as arsenic.

Iron and Silicon combine to form silicon-iron. When iron is strongly heated in contact with silica and carbon, the silica is reduced to silicon, which unites with the iron. The effect of silicon on cast iron is to set the carbon free, so that, as a rule, the greyer the pig the higher is the amount of silicon present. Silicon makes iron hard, more fusible and brittle.

The slags produced in purifying pig iron consist chiefly of basic silicate of iron ($2FeO,SiO_2$). When this silicate is heated with access of air, a very refractory substance is formed, called "bull-dog," consisting largely of oxide of iron, and an acid fusible silicate liquates out, called "bull-dog slag."

Iron and Carbon.—These two elements readily unite

together when strongly heated. The effect of carbon is to harden iron and lower its melting point. It exists in cast iron in two states—free and combined. When the carbon is chiefly in the free state as crystallized graphite or kish, the iron is called grey; when the carbon is mostly in the combined form, the iron is called white; when the carbon is present in about equal amounts of free and combined, the iron is called mottled. Iron containing carbon chiefly in the free state is said to be soft, while the combined carbon renders it harder and more brittle. Cast or pig iron is iron containing more than $1\frac{1}{2}$ per cent. carbon. When the amount of carbon is about ·25 per cent., the metal can be sensibly hardened by making it red hot and quenching in water, it is then called steel. Iron is capable of taking up and retaining about 5 per cent. of carbon, and a much greater quantity if manganese be also present. The amount of carbon usually present in cast-iron is from $2\frac{1}{2}$ to $3\frac{1}{2}$ per cent.

ALLOYS OF IRON.

Gold alloys with iron, making it harder and more fusible. Platinum when present in iron to the extent of 1 per cent., makes it fine-grained, tenacious, tough, and ductile. Copper-tin alloys unite with iron, which increases the strength of the bronze. The same remarks apply to copper zinc alloys. Copper in iron makes the iron or steel red short, and pig iron containing a notable quantity of copper is unsuitable for making wrought iron. Tin and iron form a useful alloy in tin plate. A zinc and iron alloy is formed in galvanizing.

Manganese very frequently occurs in iron ores, and is reduced along with the iron. White pig iron containing 5 to 10 per cent. of manganese is known as spiegel-eisen. When large quantities of manganese are present, the alloy is called ferro-manganese, which is hard, crystalline, and brittle; it may contain 75 to 80 per cent. of manganese. These bodies are extensively used in steel making.

Chromium alloys with iron, making it hard, white, brittle, and more fusible.

Tungstic acid, in the presence of iron and carbon at a high temperature, is decomposed, the tungsten alloying with the iron to form a hard, fine-grained white steel of great tenacity.

Properties of Iron.

Malleable iron is of a greyish white colour, having a granular, crystalline, or fibrous fracture, according to the mode of treatment. When rolled or hammered hot, the iron becomes fibrous, but continued cold hammering induces a crystalline or granular structure, making it hard and brittle. The nature of the fractured surface varies also with the manner in which the iron has been broken, for specimens broken by progressively increasing stresses are invariably fibrous, whilst the same specimen broken by a sudden blow will be crystalline. The presence of impurities generally tends to impart a granular or crystalline fracture, and makes the iron less malleable. When impurities, such as sulphur and arsenic, render the metal unworkable at a red heat, it is said to be hot or red short. On the other hand, some substances, such as phosphorus, cause iron to crack when hammered cold, it is then termed cold short. The specific gravity of iron is about 7·7 : its fusing point is said to be about 2000° C.; but before melting it assumes a pasty form, at which point two pieces may be joined together by welding. To ensure a good weld the surfaces must be clean and the metal at a white heat. In order to dissolve any scale, the smith adds a little sand, which unites with the oxide and forms a fusible silicate. When steel is being welded a mixture of borax and sal-ammoniac is preferred to dissolve the scale. The presence of any foreign bodies such as carbon, silicon, sulphur, phosphorus, copper, oxygen, etc., increases the difficulty of welding. Iron possesses considerable malleability, ductility, and tenacity. Its tensile strength ranges from 17 to 30 tons per square inch, but this like all the other physical properties is modified by the presence of impurities, which tend to make it harder, more fusible, and brittle. When iron is heated to dazzling whiteness, it burns, forming Fe_3O_4, the iron becoming friable and brittle, termed "burnt iron." Iron may be magnetized by bringing it in contact with a magnet, but it loses its magnetism when the exciting magnet is withdrawn. Its specific heat is ·1137; its conductivity about 170, silver being taken as 1000. Its electric resistance is 5·8 times that of pure copper. Iron when exposed to moist air readily rusts or oxidizes, so that it is often coated to prevent this action, as for instance, by tinning, galvanizing, and painting. Prof. Barff preserves iron from rusting by exposing it at a red heat to superheated steam, which imparts to it a coating of the black oxide (Fe_3O_4).

Preparation of Iron Ores.

In some cases poor argillaceous ores are separated from clay and sand by sifting, crushing, and washing. Clay ores occurring with shale in the coal measures may be largely separated from the shale by "weathering." In like manner pyritic ores are oxidized by the combined action of air and moisture forming soluble sulphates, which are washed away to a great extent by rain. If the weathering is pushed too far, the whole mass, especially when much limestone is present, will crumble to powder. When the sulphur in a spathic ore does not exceed ·3 to ·5 per cent., then 2 to 3 per cent. of lime and magnesia is sufficient to combine with the sulphuric acid. For regularity of working, the ore and flux should be reduced to small lumps of uniform size, which size should be proportional to the height of the furnace.

Calcination and Roasting.—All ores, except massive red hæmatite and certain magnetites, are calcined previously to reduction. This process has several advantages:—1. The amount of iron is concentrated in a less weight of matter. 2. The ore is rendered less porous and some impurities are removed. 3. Protoxide compounds of iron are oxidized to sesquioxide which does not unite with silica. Magnetite by roasting is also converted into ferric oxide (Fe_2O_3).

Roasting is performed in piles, in stalls, or in kilns. The first and second methods are adopted where fuel is cheap, but the operations are wasteful and imperfect. Roasting in kilns yields a more uniform product, with economy of fuel. One of the best forms is that of Gjers (Fig. 15), largely used in the Cleveland district. The body is of fire-brick cased with wrought iron plates. The diameter at base is about 14 feet, at the boshes about 20 feet, and at top about 18 feet. The bottom of the brickwork rests on cast iron plates, and the whole is supported on cast iron pillars. In the centre of the kiln is a cone about 8 feet in height and the same in diameter at the base.

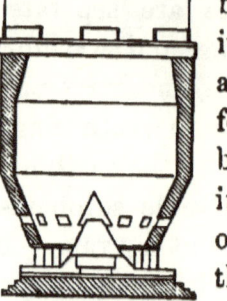

Fig. 15.

The total height is about 30 feet and the capacity about 6000 cubic feet. From two to three days are required for the roasting, with the consumption of 1 cwt. of coal per ton of ore. When the operation is complete, the ore is discharged at the bottom between the pillars, being directed outwards by the cone.

At some works the kiln is cylindrical and the heat supplied by the waste gases of the blast furnace.

Processes for the extraction of iron from its ores may be classed under two heads, viz., "Direct" and "Indirect" methods, which are represented in a general manner in the accompanying scheme (Fig. 16).

In the *direct method* the ore is reduced to iron or steel, in arrangements, such as the Catalan forge and Siemens' rotator, then hammered, re-heated and finished with the hammer or rolls. In the *indirect method*, the ore is calcined, reduced in a blast furnace, run into pig moulds, refined (or puddled direct), puddled, hammered, rolled,

EXTRACTION OF IRON. 53

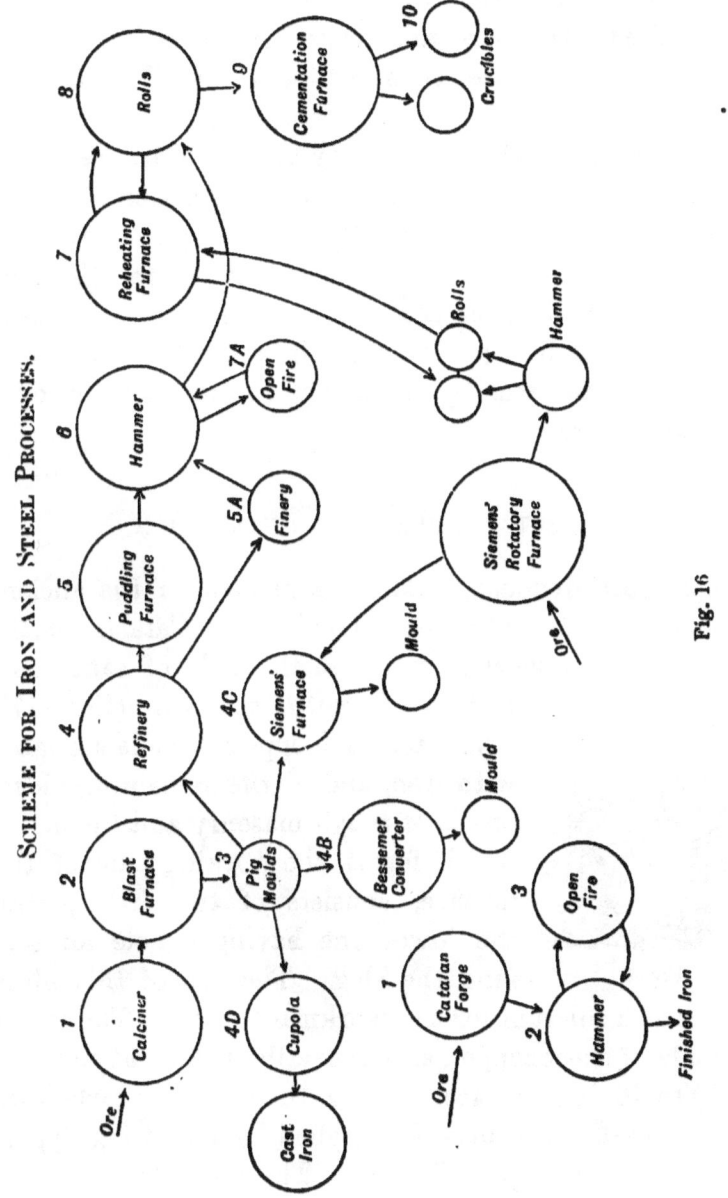

Fig. 16

re-heated, and finished in the rolls for bar iron. If steel is desired, the bars are heated with carbon in the cementation furnace, then melted in crucibles for cast steel.

In the old method, the finery and open fire were used instead of the puddling and re-heating furnaces.

Pig iron is treated in the Bessemer converter for Bessemer steel; in the Siemens' furnace for Siemens' steel; and melted in the cupola for foundry purposes. The bloom of iron produced in the rotatory furnace may be made into steel by treating it in the Siemens' open hearth furnace.

Direct Methods of Extraction.

Catalan process.—This is a survival of the ancient bloomeries still said to be carried on at Catalan in the Pyrenees. It consists of a hearth made of sandstone, and lined with charcoal (Fig. 17).

Fig. 17.

The back and opposite walls are faced with iron, and of the remaining sides, one is of rough masonry and the other, which forms the working side of the furnace, consists of two iron plates, the lower one having a hole for tapping the slag. The top of this plate serves as a fulcrum for the workman's tools. The twyer is made of sheet copper, and generally inclines at an angle of 30 to 40 degrees, according to the degree of carbonization desired. The blast is supplied by a blowing apparatus called a trompe (Fig. 18). The water falling down

CATALAN PROCESS.

the pipe A, drags air through the inclined openings BB. This aerated water falling into the cistern is divided into two streams, the air rising and passing forward into the twyer, the water running out of the cistern at C.

Hæmatites are smelted direct, but carbonates are first calcined. The ore in small pieces is charged in with charcoal, several lumps of charcoal being placed near the twyer. The heat is gradually raised until a pasty mass is formed, which is then pushed towards the twyer. After two hours the full blast is turned on and the slag tapped off. The slag is fluid and formed at the expense of the iron; it is a silicate of iron ($2FeO, SiO_2$), containing lime, magnesia, and oxide of manganese, which composition is favourable for the transference into the slag of any phosphorus that may be present. When the whole of the ore is reduced, the blast is stopped, the spongy masses of iron worked into a lump with iron tools, then carried to the shingling hammer and shingled. Then re-heated in a similar fire, and the iron finished under the hammer. If the opera-

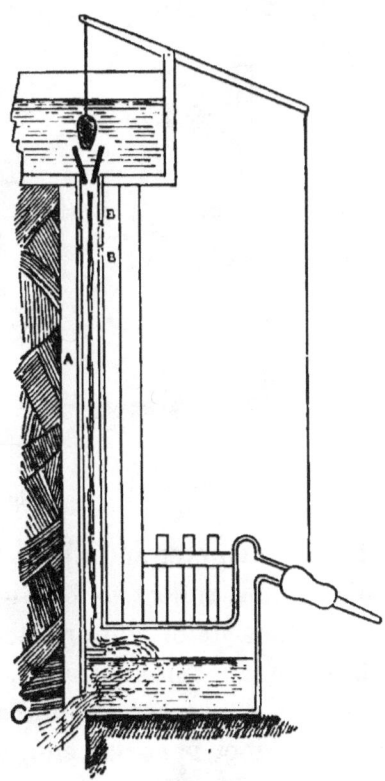

Fig. 18.

tion is prolonged and the twyer arranged at a smaller angle than in the above case, the iron combines with a certain amount of carbon and forms steel. The conditions governing the formation of slag in this hearth are similar to those of the puddling furnace and the reverse of those in the blast furnace, the former being oxidizing, the latter reducing.

Sir C. W. Siemens adapted the principle of his regenerative furnace to the direct production of iron or steel in the rotatory furnace (Fig. 19). The cylinder is

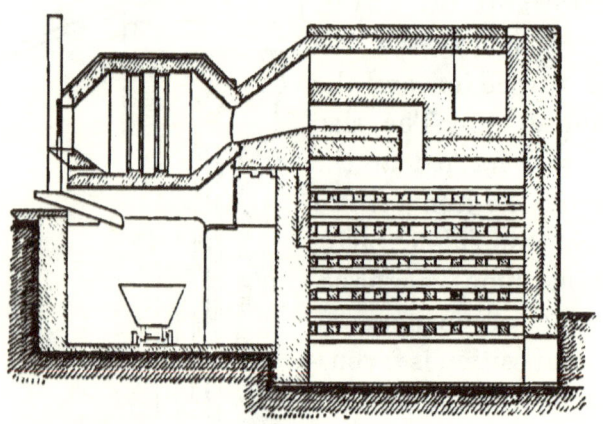

Fig. 19.

8 feet in diameter and about 9 feet long and rests on four anti-friction rollers. The inside is lined with bauxite, which consists of alumina, ferric oxide, water and a little silica. From this substance, when mixed with 3 per cent. clay and 6 per cent. plumbago, bricks are made. The working door is at one end, and beneath this is a tap hole for slag. The furnace is heated with gas supplied from a gas producer, the air being heated by passing through

one of a pair of regenerators. Ore in small pieces is charged into the hot furnace with lime, and the cylinder rotated. When red hot, about 20 per cent. of small coal is added, and the velocity increased. When the reduction is complete, the slag is tapped off; then a quick rotation is imparted to collect the iron into three or more balls, which is effected by means of projecting ribs on the interior surface of the rotator. The balls are then shingled, re-heated, and finished in the usual way. If steel is desired, 10 per cent. of spiegel-eisen is added after tapping off the slag. The charge is then melted and cast into moulds. In some cases the balls are treated for steel in the "open hearth" furnace. The advantages claimed for this method are economy of time, saving of fuel and purity of iron, since the metal is not brought in contact with solid fuel as in the blast furnace.

Indirect Method of Extraction.

The calcined ore is put into the blast furnace with coal, coke, or charcoal and a suitable flux, which is usually lime. The heat is well utilized and the reduction more perfect than in any other form of furnace. In the blast furnace there are two currents travelling in opposite directions, and constantly acting on each other—a "gaseous" ascending current and a "solid" descending one. The former travels at the rate of about 3 feet per second, the later at the rate of 3 feet per hour. The effect of the blast on the carbon of the fuel is to produce carbonic acid (CO_2) at the level of the twyers with evolution of great heat. This gas ascends, and is reduced by carbon at a very short distance from the twyers thus, $CO_2 + C = 2CO$. This carbonic oxide is the principal reducing agent in the blast furnace, the oxide of iron being reduced to the metallic state as a spongy mass, thus—

$$Fe_2O_3 + 3CO = 3CO_2 + Fe_2$$

At the same time the flux and earthy matter of the ore unite to form slag, which descends with the iron; the latter in contact with highly heated carbon is first carburized, then melts and collects in the hearth combined with other substances, such as silicon, phosphorus, sulphur, and manganese, which have also been reduced, constituting cast iron. On the top of the molten iron floats the liquid slag. The temperature and pressure have a great influence on the reducing action, and as the temperature of the furnace increases with the temperature and pressure of the blast, it follows that as the reducing energy becomes greater the metal is more impure. When very pure iron is desired, then rich hæmatite and cold blast are used, with charcoal as fuel. Great advantage is gained by the use of hot air, as less carbon is required for reduction and fusion. It is also useful to remedy defects and regulate the passage of materials in the furnace. If the fusion or reduction is at fault the temperature of the blast is raised, or more carbon is added to the furnace. The former acts promptly, while the latter often takes several hours to remedy the defect.

The quality of the pig iron produced from a given furnace will depend on the temperature, the nature of the charge, and the mode of working. With easily reducible ores and heavy burdens—that is, with a large proportion of ore to fuel—the iron will be white, since the metal is kept only the minimum time in contact with incandescent carbon. With a high temperature and a light burden the pig iron is more siliceous and grey. The same things influence the character of the slag. Blast furnace slags are double silicates of lime and alumina, and may be represented by the formula

$$3(CaO, SiO_2) + Al_2O_3, 3SiO_2$$

or

$$6(2CaO, SiO_2) + 2Al_2O_3, 3SiO_2$$

The former is the kind of slag obtained from charcoal furnaces, and the latter from furnaces using coke or coal. In both cases, the lime is replaced more or less by magnesia, oxide of iron, and oxide of manganese, while the silica is sometimes replaced to a small extent by alumina. The colour varies from white to grey, sometimes with varying shades of yellow, green, blue, and black, according to the metallic oxides present. Generally a white or grey slag accompanies grey iron, and a dark coloured slag, white iron. The former often contains excess of lime, which diminishes its fusibility; the latter is more fusible and contains oxide of iron, which, when present in quantity, makes a very liquid "scouring" slag or scoria. When forge or mill cinders are added to the charge the resulting metal is called cinder pig iron, and the change may be represented by the following equation—

$$6(2FeO, SiO_2) + 4C = 2(2FeO, 3SiO_2) + 4CO_2 + 4Fe_2$$

The "scouring" slag produced sometimes contains as

much as 20 per cent. of iron. When phosphoric acid is present in a blast furnace it is reduced, and the phosphorus passes into the iron, which can only be prevented by leaving much oxide of iron in the slag.

The modern blast furnace (Fig. 20) is an elongated barrel-shaped structure, the height being from four to five times that of the greatest width, called a "cupola." The body is formed of wrought iron plates, $\frac{1}{2}$ inch thick, riveted together, and within which is built the outer casing of ordinary masonry, the inside being lined with fire-brick about 18 inches thick, while between the two layers of brickwork is a small space filled with sand to allow for expansion and contraction. The body or stack is supported on a cast iron ring resting on iron columns, and the lower part, from the top of the columns to the tymph arch, is also cased with iron. The hearth is independent of the masonry of the stack, and is built in after the stack is completed. It requires to be made of very refractory material of considerable thickness, having to withstand a very great heat, in addition to the corrosive action of the molten slags. In Fig. 20, A is the charging gallery, B the cup and cone arrangement for charging, C the throat, D the body, EE the boshes, F the blast main, G the iron ring

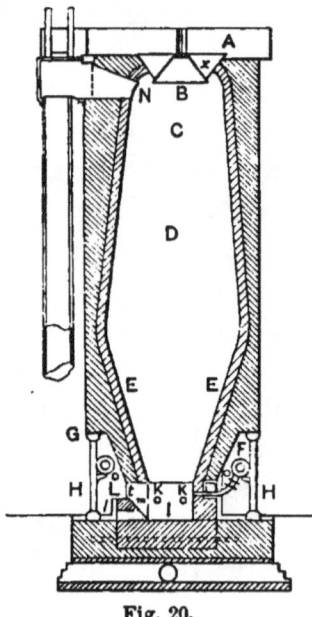

Fig. 20.

supporting the body, HH the pillars, I the hearth, K the twyers, L the dam, l the iron dam plate, m the fore-hearth. The dam is formed of fire-brick, and is carried up to the twyer level, a semi-circular notch in the top edge serving as a passage for the slag. The tap hole for the molten iron is a narrow slit through the bottom of the dam. The arch covering the fore-hearth is called the tymph arch (t). N is the opening for collecting the waste gases which are utilized for heating the blast, boilers, etc.

The charge is tipped into the cup or hopper (x), and allowed to fall into the furnace by lowering the cone B, which acts very advantageously in distributing the charge over the surface of the materials already in the furnace. The hot blast twyer consists of a hollow conical wrought iron pipe, with double walls for the circulation of a constant stream of water and a central pipe connected with the blast main and inserted into the above conical pipe.

The arrangements for heating the blast are of two kinds—cast iron pipes through which the air passes, heated externally, and chambers of refractory brickwork constructed on the principle of Siemens' regenerators, which are now in most general use. The use of the hot blast was patented by Neilson in 1828, and first adopted at the Clyde Iron Works for the blast furnace. It is now almost universally adopted.

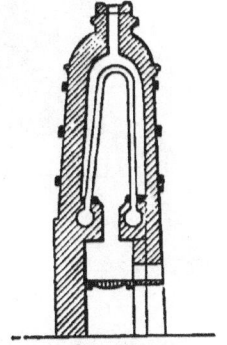

Fig. 21.

Cast iron stoves.—Fig. 21 represents an old form of

stove, which consists of a chamber containing a series of arched pipes of cast iron arranged in an oblong fire-brick chamber, along each of the long sides of which are two circular mains fitted with sockets into which the legs of the vertical pipes are received, while between the mains, and running the full length of the stove, is a rectangular fireplace.

Of the "regenerative" type two principal forms are employed, invented respectively by Cowper and Whitwell.

The Cowper stove (Fig. 21a) is a circular wrought iron tower, closed with a dome-shaped roof, lined internally with fire-brick. It contains a circular fire-brick flame flue A, into which the waste gases from the blast furnace pass by the valve B. The body of the stove is occupied by a chequer work of fire-brick for absorbing the heat. The gas entering the flue A is there burnt, the necessary air for combustion entering by the valve C. The hot products passing down through the chequer work make it red hot, and finally pass into the chimney flue D.

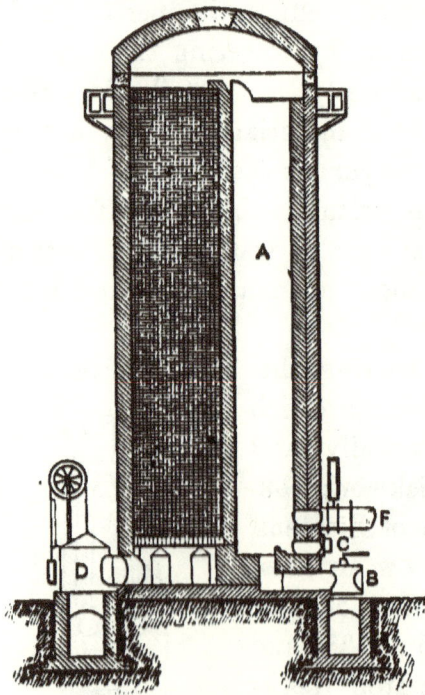

Fig. 21 a.

WHITWELL'S STOVE.

The stove having been thus heated, the valves B, C, and D are closed, and the cold blast valve at the bottom of the stove opened, as well as the hot blast valve F. The cold air enters at the bottom or cooler end, and ascends through the brickwork, getting gradually hotter and then escapes by the valve F, which communicates with the blast furnace, at a temperature of about 800° C. Two stoves are worked in conjunction, one being heated by the combustion of the waste gases, while the other is being utilized in heating the blast. These stoves are 50 to 55 feet high, and 20 to 25 feet in diameter.

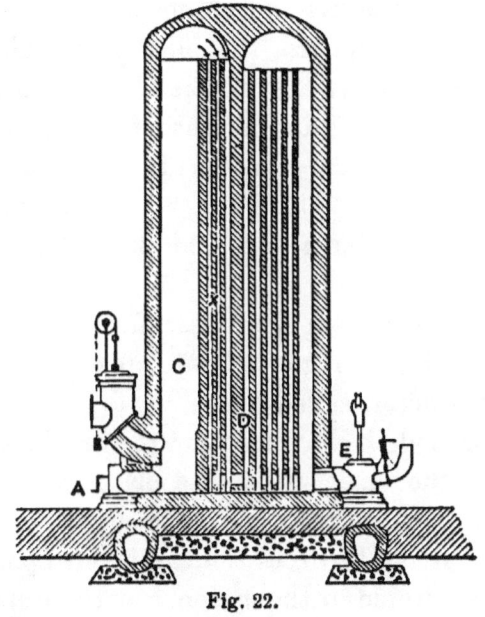

Fig. 22.

Whitwell's stove differs from Cowper's, chiefly in the arrangement of the absorbing brickwork. The waste gases and air for combustion are admitted at several points of the stove, so that the combustion is more perfect. The regenerative brickwork is built so as to form a number of long and narrow vertical chambers, communicating with each other at the top and bottom. The blast furnace gases enter through A, and meeting with warm air introduced through suitable passages, combustion takes place, and the flame rising up C, passes down the narrow passages, x, to the bottom. Here more air is admitted to burn the unconsumed gases, which rise up D and down through another series of narrow channels, leaving at the chimney valve E. When the stove is sufficiently heated, which takes about two hours, the gas and chimney valves are closed and the blast valve opened, the blast passing in inverse order to that of the gases, as in the Cowper stove.

The waste gases from a blast furnace contain by volume from 55 to 60 per cent. nitrogen, 23 to 28 per cent. carbonic oxide, 7 to 12 per cent. carbonic acid, about 2 per cent. marsh gas, and 2 to 6 per cent. of hydrogen.

The iron tapped from the blast furnace is run into sand moulds, forming ingots about 3 feet long and 3 or 4 inches in diameter, called pigs.

Various methods have from time to time been devised for collecting the waste gases of a blast furnace besides the cup and cone contrivance already described. In some small charcoal furnaces the top is left open, and an annular space formed in the masonry at the upper end of the body of the furnace. This space communicates with the interior by a number of openings inclining upwards beneath the level of the charge, through which some of the waste gases are drawn. In some cases a wrought iron cylinder is suspended in the mouth of the furnace; in others a brick tube is built in the throat, being supported by arms of brickwork. Langen withdraws the gases from the centre, instead of the sides, by suspending a bell above the throat, which bell can be raised by a lever for charging, the joints being kept tight by water.

Sometimes the charge of a furnace is obstructed in its descent; the lower portion being melted and withdrawn, leaves a "scaffold," which, with the increasing weight from above, sometimes suddenly gives way and falls into the hearth. This is called a "slip," and deranges the working of the furnace.

In some cases, large ferruginous masses, called "bears," are formed in the hearth of a blast furnace, often containing iron, copper, silicon, graphite, manganese, nickel, cobalt and cyano-nitride of titanium, which is a compound resembling copper in colour.

Refining Pig Iron.

For the better qualities of wrought iron, crude pig iron was formerly submitted to a preliminary operation in a rectangular hearth, called a refinery (Fig. 23). A number of twyers are so arranged on two of its sides as to project a stream of air on to the molten iron, to oxidize its impurities. The iron is finally quenched by running it into an iron trough surrounded with water, thus causing the carbon to remain in the combined form, which condition is necessary for puddling by the original or "dry" method, as white pig iron assumes a pasty state when strongly heated.

The element whose chemical affinity for oxygen is greatest will absorb this gas first, and others will be oxidized in turn in proportion to their chemical energy. The order of oxidation will be as follows :— Silicon, manganese, phosphorus, sulphur, carbon and iron. But oxygen will combine with the dominant metal by the influence of mass, so that refining slags always contain a large amount of oxide of iron.

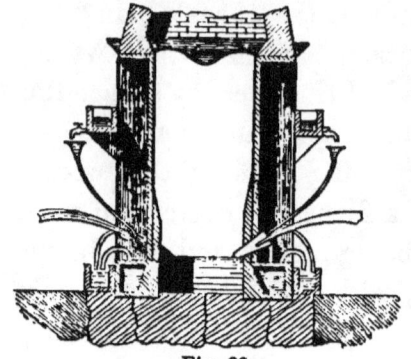

Fig. 23.

Puddling—Dry and Wet.

The method of "dry puddling" in a reverberatory furnace was developed by Cort, and patented by him in 1784. Each operation is composed of three periods—fusion, rabbling, and forming the blooms. White or refined iron is chiefly used. About 4 cwts. of metal is charged into the furnace along with rich slags, and is partially melted in about half an hour, forming a pasty mass. It is then stirred with iron tools so as to bring all parts under the oxidizing influence of the air. As the impurities are removed, the iron becomes less fusible, requiring the temperature to be gradually raised. The particles of iron are then collected into balls by the puddler, each ball weighing about 80 lbs.

In "wet puddling," which has superseded the dry method, the preliminary refining is dispensed with. It is characterized by the complete fluidity of the iron and the greater length of time required to remove the impurities. The pigs submitted to wet puddling are siliceous or strongly carburized, and should contain manganese, if steel is desired. The action of manganese in puddling tends to retard the decarburization of the iron by reducing Fe_3O_4 to FeO, as carbon is especially removed by the oxidizing action of Fe_3O_4, so that as long as the iron contains manganese, its strong affinity for oxygen prevents Fe_3O_4 being formed. Also slags containing oxide of manganese are more fluid and therefore more easily expelled from the iron by hammering and re-heating.

The bed and sides of a modern puddling furnace are lined with refractory materials rich in oxide of iron. When the iron is melted, it is between two oxidizing influences—the air and oxide of iron—and the operation will be shorter and the product more uniform as the rabbling (or moving of the mass by the tools of the puddler) is more vigorous. The ferruginous slag formed, takes up oxygen from the air, causing FeO to pass into Fe_3O_4, which then oxidizes the impurities in the order of their oxidizability, viz., silicon, manganese, phosphorus, sulphur, and carbon.

The period while the carbon is being oxidized, forming carbonic oxide, is called the "boiling" stage. The whole mass is then in a state of violent agitation, the slag being lifted up some distance above the bed, often flowing out at the door in a liquid state. The iron gradually comes to nature and is collected in balls as before, ready for the hammer.

The slag from a puddling furnace is essentially a silicate of iron ($2FeO, SiO_2$), containing many of the impurities originally present in the iron, and called "tap cinder."

The puddling furnace (Fig. 24) is a reverberatory, with a low flat roof slanting from fireplace to flue. The fire bridge A and flue bridge B are formed of hollow iron castings encased in fire-brick, and the bed is likewise formed of iron plates rebatted together. The sides generally consist of hollow iron castings, which can be kept cool by the circulation of air through them. The laboratory part C is about 6 feet long and 4 feet wide, tapering to the flue bridge. The grate area varies from one-third to one-half that of the laboratory. The bed is lined with broken slags, hammer scale, and red oxide of iron, and the sides with bull dog, all being well rammed down. This is termed "fettling." The whole brickwork is cased with side plates of cast iron, united by flanges and bolts, and bound together across the top with wrought iron tie-rods.

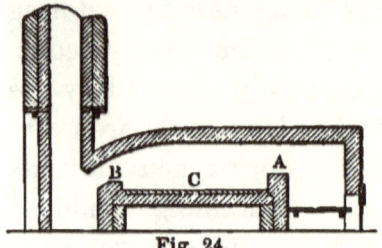
Fig. 24.

Treatment of puddled balls.—The operations are conducted in that part of ironworks known as the "forge," which includes puddling, shingling, rolling, etc. The oldest class of hammers for shingling are the "tilt," where the axis is between the point of application of the cam and the head, and the "helve," or lift hammer, in which the hammer block and lifting cam are on the same side of the fulcrum. The former is used for light work, the latter for heavier work, such as shingling puddled balls, blooming piles, etc. The modern form is the steam hammer, employed both for shingling and welding.

Various forms of squeezers are also used instead of the hammer for the treatment of puddled balls.

The rolling mill is generally used for merchant iron, the rolls being of two kinds—"roughing" rolls and "finishing" rolls of several forms, according to the shape of the bar required.

Re-heating.—The shingled masses and bars of iron obtained from the forge contain slag and impurities which have not been eliminated in the previous operations and the object of re-heating is to remove these substances as much as possible. The bars are cut into short lengths, made into rectangular piles, bound together with a strip of iron and raised to a white heat. The enclosed slag liquates out as a fluid basic silicate ($2FeO, SiO_2$), leaving a basic residue mixed with black oxide of iron (Fe_3O_4), which is further removed in rolling. The basic slag also acts on the manganese, silicon, phosphorus, etc. oxidizing them, the oxides passing into the fluid slag. The sulphur and carbon are also oxidized in the same way, and removed as gases.

The re-heating furnace (Fig. 25)

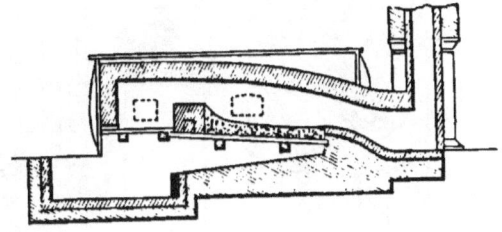

Fig. 25.

is somewhat similar in shape to the puddling furnace; the bed is flat and slightly inclines to the flue and is lined with sand. In some works basic material is used.

For merchant bar, the iron bars are cut up, made into small bundles called fagoting, raised to a white heat, welded by the hammer, and rolled. This produces a stronger iron. If the operation is again repeated, the iron is still further improved, and called " treble " best.

Of late years gas furnaces have been used for re-heating, such as Siemens' furnace, Fig. 8. A modification of the regenerative furnace has been applied to re-heating by Ponsard. In this arrangement the gas from the producer is delivered directly to the furnace and burnt with hot air, heated by a "recuperator" placed under the bed of the furnace. The recuperator is a brick chamber, partly solid and partly hollow, containing a number of vertical passages, the adjacent ones being separated, and the

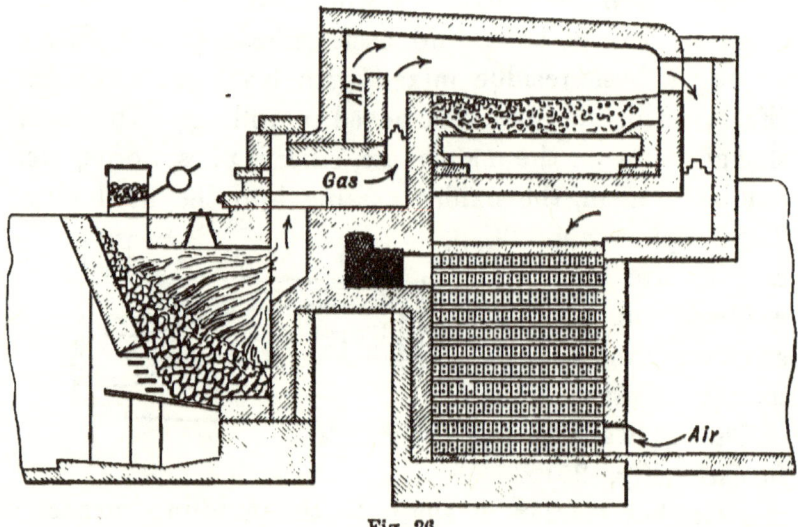

Fig. 26.

alternate ones connected by horizontal passages in the perforated bricks. The flame passes downwards, while the cold air, admitted at the bottom, passes upwards by separate channels, the action being continuous. The whole arrangement is seen in Fig. 26.

That part of the works containing the re-heating furnaces, and the appliances connected therewith, is termed the "mill."

Conversion of Pig Iron into Malleable Iron in Open Fires.

South Wales finery.—In South Wales and other places a superior quality of iron sheets for tin plates is made from pig iron in the "finery," or open fire (Fig. 27). The hearth is rectangular, and formed of cast iron plates lined with charcoal dust, the bottom being kept cool by a current of air. Three sides are vertical, while the remaining side slopes a little outwards. The fuel employed is charcoal, the fire being blown by a blast through a single twyer. The operation is analogous to dry puddling. The charge of $2\frac{1}{2}$ to 3 cwts. of refined iron from a coke "refinery" (Fig. 23) produces a finery ball, weighing about 2 cwts., which is shingled and drawn out to a long bar, 2 inches thick, under a lever hammer. The bar is then nicked and broken into pieces, the best pieces selected, piled, and re-heated in the flame of a coke fire, in a furnace similar to the finery, known as the "hollow fire," the upper part of which forms a chamber in which

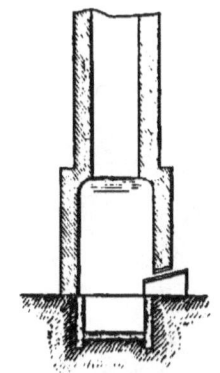

Fig. 27.

the piles are re-heated. The piles are then welded under the hammer and rolled into sheets. Before finishing, the sheets are pickled in sulphuric acid, and then rolled cold. Finery slag is highly basic, containing upwards of 75 per cent. of FeO.

The Swedish-Lancashire finery is arched over at the top, and communicates with the chimney by a horizontal flue in which the pig iron undergoes a preliminary heating

CHAPTER V.

STEEL.

EVERY malleable sample of iron containing sufficient carbon to enable it to be hardened is called "steel," the lowest carbon limit being ·15 per cent. When the carbon exceeds 1·8 per cent. it forms cast iron. When the steel has been produced in the molten state and poured into an ingot, as in the Bessemer process, it is called "ingot" or cast steel; when produced in the solid state, as by puddling, it is called "wrought" or weld steel.

Steel has a light bluish grey colour, which becomes whiter after hardening; the lustre is similar to that of iron; the fractured surface presents a fine uniformly granular texture, being very close grained when hard. Best steel has a higher tenacity than any other metal. It is hardened by making it red hot and plunging it into water, which slightly increases its volume. Hard steel is softened by making it red hot and cooling it slowly. The melting point is between that of cast iron and wrought iron, being lower the greater the amount of carbon it contains. When the amount of carbon is less than ·4 per cent., it is known as mild or soft steel.

Effect of impurities.—·5 per cent. copper makes steel

red short. Chromium gives whiteness and brilliancy to steel, but makes it harder and more brittle. Phosphorus is much more injurious in steel than in iron, and steel containing it can only be worked by keeping the carbon low. Manganese hardens iron like carbon, but in a less degree; it also prevents the separation of carbon in the form of graphite, which is the opposite of silicon. Silicon hardens steel, makes it brittle and less tenacious. Sulphur has much the same effect, and makes steel red short.

Steel is produced (1) by direct methods as in the Catalan forge; (2) from pig iron in the finery; (3) by puddling pig iron; (4) by the cementation process; (5) by melting raw steel in crucibles; (6) by the pneumatic process as in the Bessemer vessel; (7) in the open hearth, as in Siemens' process.

The method of making steel in the Catalan forge and Siemens' rotatory furnace has been already indicated.

When a "finery" is used for making steel, the hearth is shallower than that used in making wrought iron, the twyer is fixed at a lower level and at a greater inclination, more fuel is used, and half as much more time required. The best varieties of iron for the purpose are strongly mottled pig iron and spiegel-eisen. The iron is purified by the oxidizing action of the blast and slag, and carburized by the charcoal used as fuel.

Puddled steel is produced in much the same way as iron, but greater care is exercised in selecting the pig irons used, those richest in carbon and manganese being preferred. A higher temperature is required than for iron, so as to form a thin slag, which covers the iron and retards the decarburization. The balling is done at as low a temperature as possible, to prevent the carbon burning off. The mechanical treatment of the puddled balls is similar to that for iron.

Cementation process.—This consists of exposing bars of malleable iron in contact with charcoal to a high and prolonged temperature in closed vessels, from which the air is excluded, in this way forming blister steel. The furnace (Fig. 28) is rectangular in plan, and covered with an arch,

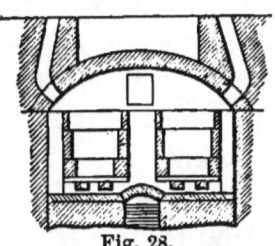

Fig. 28.

having a hole in the centre 12 inches square, which is opened during cooling. It contains two rectangular fire-brick vessels, arranged one on each side the fireplace, which runs the whole length of the furnace. Each of these vessels is about 12 feet long and 3 feet square, but the sizes differ in different works.

Swedish bar iron is preferred for making the best kinds of steel, and hammered bars are generally preferred to rolled bars. Ordinary sizes of bars are 2 to 5 inches wide, and $\frac{3}{4}$ inch thick. The pots are charged by covering the bottom with coarse charcoal, and on this placing a row of iron bars, and so on alternately until the vessel is full. The top is then covered with "wheel swarf" from grindstones, which melts and forms a glaze when heated, thus excluding the air. The charge for one pot may be from 10 to 30 tons of iron. Two small holes are left in the furnace, corresponding to two similar holes in the vessels through which a bar from each vessel projects, called the trial bar. When the fire is lighted, two to three days are required to attain a proper temperature, which is then maintained for seven to nine days, according to the temper of steel required, the hardest steel requiring the longest time. On examining the trial bar from time to time the conversion is seen to have penetrated gradually to the centre, when the vessels are allowed six days to cool down. By cementation, the fibrous iron is changed to a finely granular steel, and the surface of each bar is covered with blisters, which in good steel are small and fairly regular. These blisters are probably due to the reduction of oxide of iron contained in the enclosed slag, and the evolution of carbonic oxide. If air has gained access to the bars during conversion, they are rough and have a skin of iron. When the temperature has been too high, and fusion on the surface has taken place, they are called "glazed" bars, both being unfit for making cast steel.

Blister steel is used without further treatment for common purposes, but generally the bars are fagotted and welded one or more times, forming shear and double shear steel. The texture of the steel is modified accordingly as it has been rolled or hammered. The particles of rolled steel are somewhat spherical, such steel combining great elasticity with toughness. Hammered steel has a finer grain, a more compact structure, and a greater uniformity and density than rolled steel.

Case hardening.—This consists of the production of a surface coating of steel by a short cementation. Small articles may be case hardened by rubbing them when red hot in powdered yellow prussiate of potash (K_4FeCy_6), and when the substance has volatilized, plunging them into water.

Cast steel.—This was introduced by Huntsman in 1740, who succeeded in melting blister steel in covered crucibles, and pouring the metal into moulds, thus obtaining a more homogeneous product than by hammering or rolling. In 1801, Mushet introduced the method of making cast steel by melting malleable scrap iron with charcoal and black oxide of manganese (MnO_2) in crucibles, which has since been largely practised.

Indian or Wootz steel is made by melting 1 lb. of malleable iron in a small crucible, with 10 per cent. of dried wood and 2 or 3 large leaves, producing a very hard steel.

Open hearth steel.—This class of steel is made in the Siemens' type of gas furnace, known as the "open hearth," and the methods pursued may be classified under three heads—(1) Pig and scrap process; (2) pig and ore process; (3) pig and blooms process. The first is due to Martin, the two last to Siemens, and known as the Martin and Siemens processes respectively. At the present time a combination of No. 1 and No. 2 is most commonly employed, and termed the Siemens-Martin process.

The Martin process consists of melting malleable iron with pig iron, the latter with low silicon being preferable. The cast iron is melted in a separate furnace, and run into the furnace in the fluid state, then the malleable scrap (previously made red hot) is added. The oxide of iron (Fe_3O_4) re-acts on the carbon and other elements of the cast iron producing gases which escape and agitate the molten metal, tending to make the iron uniform. When the refining is completed, spiegel-eisen or ferro-manganese is added, accordingly as hard or mild steel may be desired, and the metal tapped.

In the Siemens-Martin process, the scrap iron is partly replaced by pure rich oxides of iron, which assist the atmosphere in oxidizing the impurities, thus hastening the process and rendering it less costly. The pig and scrap are added cold to the hot furnace, and when the ebullition due to the action of the scrap on the cast iron has nearly ceased, about 15 per cent. of ore is added. When the purification is complete, 8 to 12 per cent. of spiegel-eisen or ferro-manganese is added, and the metal tapped as before.

In the third method mentioned above the balls or blooms are furnished by Siemens rotatory furnace or by the puddling furnace. These blooms act on the cast iron like the scrap and ore in the last process.

In the ordinary Siemens furnace, the oxidizing action is feeble because of the siliceous nature of the slag; the removal of sulphur and phosphorus being almost nil. Of late years these elements have been largely removed by having a basic lining to the furnace, such as oxide of iron or dolomite, instead of the original sand lining.

Pernot has modified the Siemens furnace by having a movable bed which inclines at an angle of 5 to 6 degrees. The furnace (Fig. 29) is circular and cased with wrought iron, the roof being fixed. It is heated with gas in the usual way. The bed is lined with basic material rich in oxide of iron, and supported by a central spindle and conical friction rollers on a travelling carriage. By means of geared wheels it is slowly rotated, making three to four revolutions per minute during charging and working. This rotation not only agitates the liquid metal, but favours the re-action of the basic lining on the cast iron. By this rotation and inclination of the bed, each half is alternately exposed to the flame, so that the bottom heat is constantly renewed, and sticking of the charge prevented; also the FeO of the fettling is oxidized to Fe_3O_4, which is brought under the liquid metal, and oxidizes the impurities together with some of the iron. The charge consists of $\frac{1}{5}$th pig iron and $\frac{4}{5}$th scrap wrought iron and steel. Spiegel-eisen or ferro-manganese is added at the end as usual.

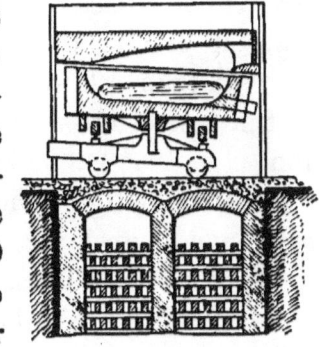

Fig. 29.

Ponsard adds a wind chest to the central spindle, which is hollow and communicates by a pipe with twyers fixed in the side of the furnace, so that an oxidizing blast may be injected as in the Bessemer process. The bed in this case only moves through half a revolution, otherwise the twyers would be alternately in and out of the metal.

Malleable cast iron.—The method of rendering cast iron articles malleable is effected by an oxidizing cementation; hæmatite, lime, or other metallic oxides being used as the cement. The pig iron used for casting must be of a superior quality, charcoal iron being preferred.

The castings are packed in chests with alternate layers, generally of Cumberland hæmatite. The temperature is raised gradually, and maintained at a red heat for one to three days, according to the softness desired; then allowed a few days to cool. The effect of the hæmatite is to oxidize the carbon of the pig iron.

Steel casting.—The great difficulty in casting steel is to avoid blowholes. A ladle of Bessemer or open hearth steel is a seething mass containing oxygen, and in this condition is unfit for castings.

Various methods are employed to produce sound castings. A triple compound of iron, manganese, and silicon, is commonly used. The manganese and silicon reduce the Fe_2O_3, forming silicate of iron and manganese, which, being very fluid, liquates out from the solidified steel.

Blowholes are believed to be due to the escape of hydrogen and carbonic oxide, the latter being formed by the re-action between the slag and carbon present, so that silicon and manganese by their superior oxidizability over carbon, prevent carbonic oxide being formed. Large castings require annealing in order to equalize the strain in all parts.

BESSEMER PROCESS.

This consists of blowing air through molten pig iron, without the aid of ordinary fuel, in a vessel called a converter, whereby the carbon, silicon, and some of the iron are oxidized, producing a very high temperature and leaving the iron commercially pure.

Two distinct modes of working, employing totally different classes of pig iron are now adopted. In one the converter is lined with an acid material called ganister, and may be designated the "Acid" Bessemer process; in the other, the vessel is lined with calcined dolomite, which is a basic material. This is called the "Basic" Bessemer process. In the acid process, the iron employed must be free from sulphur and phosphorus, as these elements cannot be removed from the iron, so that only the purer classes of pig iron can be used, such as those obtained from Cumberland hæmatite.

The converter is sometimes used fixed, as in Clapp-Griffith's process, but more generally it is arranged so as to be moved through an angle of 180°, thus enabling the metal and slag to be poured from the mouth. Figs. 30 and 31 represent the most usual form of vessel; it consists of a shell of wrought iron plates riveted together, the neck being inclined at an angle of 30° to the body. The centre of the body is enclosed with a stout band of iron, upon which are fixed two arms, called trunnions, by which the vessel is suspended on iron standards. One of the trunnions is hollow, through which the blast of air passes and thence through a pipe to the twyer box, which forms the movable bottom of the converter, and is kept in position by bolts and cotters. The bottom of the vessel is perforated by 10 to 15 circular holes, into each of which is placed a conical fire-clay twyer 20 to 22 inches long, and perforated with 10 or 12 holes, each $\frac{3}{8}$th inch in diameter. The lining of the converter is of ganister,

Fig. 30.

F

9 to 12 inches thick, which contains 85 to 90 per cent. of silica. The ganister is coarsely ground, mixed with a little powdered fire-clay and water, then well rammed in between the iron shell and a central wooden core having the internal form of the converter.

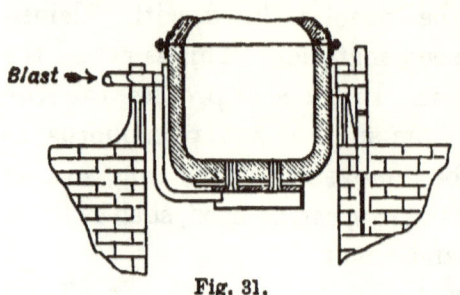

Fig. 31.

The newest form of converter is concentric, the neck being straight, so that it may be charged or emptied from either side, which allows the wear of the lining to be equalized.

The lining having been dried, the converter is made red hot by burning coke inside, which is then tipped out, and the vessel brought into the horizontal position. Molten pig iron is then run in from a cupola or reverberatory melting furnace, in amount varying from $1\frac{1}{2}$ to 10 tons, according to the size of the converter. It is then rotated into the vertical position, the blast being turned on so as to prevent the metal running through the twyer holes. For the first five minutes the flame is but slightly luminous, and but few sparks are emitted. During this period the graphite is converted into combined carbon, and the silicon is oxidized, forming a silicate of iron and manganese; then the action gradually becomes very violent, the flame is brilliant, and showers of sparks, consisting of carbon, iron and slag are thrown out in consequence of the rapid ebullition, produced by the escape of carbonic oxide from all parts of the metal.

This lasts for seven or eight minutes, when the action diminishes. When the last trace of carbon is burned, the flame suddenly drops, and if the blast is continued, the iron itself burns. The converter is then brought to the horizontal position, and about 10 per cent. of spiegel-eisen, or an equivalent amount of ferro-manganese, is run in, or dropped in solid. The metal is then poured into the ladle, and from thence into ingot moulds. One blow lasts about twenty minutes. As iron, after decarburization in the Bessemer vessel always contains oxygen, the object of adding spiegel-eisen or ferro-manganese is to give sufficient manganese to combine with this oxygen and form a fusible slag with silica, and enough carbon to convert the iron into steel. The slag from the acid Bessemer process contains excess of silica and thus differs from tap cinder which is neutral or basic.

Basic process.—The application of a basic lining to the converter was introduced by Thomas and Gilchrist, thus enabling common pig iron, containing much phosphorus, to be treated. The converter generally used in the basic process is shorter and wider than the acid-lined one. The lining consists of ground dolomite mixed with anhydrous tar as a cement. In the acid process, grey iron rich in silicon is necessary, while in the basic process white iron with less than 1 per cent. of silicon and 2 to 3 per cent. of phosphorus is employed, the combustion of the latter furnishing the necessary temperature. The lining of the basic vessel is much sooner worn away, and the end of the operation more difficult to determine than in the acid-lined converter, so that samples have to be taken and tested, by their mode of fracture and the appearance of the grain. The operation in the basic vessel is conducted in the same way as in the acid process until the carbon is burnt off; then the blowing is continued to remove the phosphorus, which is called the "after blow." The spiegel-eisen is then added as before, and the metal tested.

CHAPTER VI.

SILVER.

SILVER is remarkable for its whiteness and brilliant lustre, although when precipitated from its solutions it often forms a grey powder; it is harder than gold but softer than copper, the relative hardness being as 4 : 5 : 7·2. Silver is extremely malleable and ductile, with a tenacity of about 14 tons per square inch; its specific gravity is 10·5, which may be slightly increased by coining, rolling, hammering, etc.; it melts at about 1000° C.; is one of the best conductors of heat and electricity; is volatile at high temperatures, and at the temperature of the electric arc may be boiled and distilled. When heated in a current of hydrogen it volatilizes at 1330° C. It does not oxidize when heated in air, but molten silver mechanically absorbs oxygen and emits it on solidifying; this is called "spitting." Silver in a finely divided state is oxidized when heated with certain metallic oxides, such as CuO, MnO_2, Pb_3O_4, etc., these bodies being reduced to lower oxides. Silver is soluble in nitric and sulphuric acids. Silver unites readily with sulphur when heated, forming silver sulphide (Ag_2S), which is a dark grey, crystalline body, with feeble lustre; somewhat soft and malleable. When heated in air it does not form oxide or sulphate like most other metallic sulphides and at a red heat is decomposed into metallic silver and sulphurous acid. If, however, silver sulphide mixed with copper or iron sulphide is roasted, then silver sulphate is formed, which is soluble in water. Dilute hydrochloric acid has no action on silver sulphide, but the strong acid attacks it. Lead, copper, or iron decompose it when the two bodies are fused together. When silver sulphide is heated with common salt, in the presence of moist air, silver chloride is formed.

Silver, and all its salts dissolve in sodium hyposulphite, forming a soluble double hyposulphite ($Na_2S_2O_3$, + $Ag_2S_2O_3$), when the sodium salt is in excess. Silver combines directly with chlorine to form silver chloride ($AgCl$). The same substance is formed by adding hydrochloric acid or a solution of common salt to a solution containing silver, when AgCl is precipitated as a white powder; if, however, a large excess of strong salt solution be used, the AgCl is dissolved, a double salt being formed thus :—

$$AgNO_3 + NaCl = AgCl + NaNO_3$$
$$AgCl + NaCl = (AgCl, NaCl).$$

Silver chloride fuses at a low red heat to a yellow liquid and readily volatilizes at a strong red heat. It is insoluble in acids, but soluble in ammonia, sodium chloride, sodium hyposulphite, and potassium cyanide. It may be reduced by hydrogen, carbonate of soda, zinc, iron, and several other metals, and partially by sulphur. It unites with oxide of lead in all proportions, and partially so with sulphide of lead and some other sulphides.

EXTRACTION OF SILVER.

Alloys of silver.—Silver forms valuable alloys with copper, but most of them undergo a partial liquation on cooling. When the copper is under 50 per cent. the alloys are white, but when above that amount, they have a red tint. The alloy containing 630·29 of silver per 1000 is more fusible than silver or any alloy of silver and copper. Standard silver, used for the British coinage contains 92·5 per cent. of silver, and 7·5 per cent. copper. Silver solders consist of silver and copper with a little brass; sometimes bronze and arsenic are added.

Ores of silver.—Silver occurs native; as sulphide in silver-glance; as chloride in horn silver; as bromide and iodide; also in many lead, zinc and copper ores, and sometimes in iron pyrites.

METHODS OF EXTRACTION.

Silver is extracted from its ores by both "Wet" and "Dry" methods, which may be classified as follows— 1. Liquation; 2. Amalgamation; 3. Lead method; 4. Wet methods.

LIQUATION.

Liquation is employed to separate silver from argentiferous copper by dissolving it in molten lead, and consists of three processes—1. Alloying with lead; 2. Liquation proper; 3. Treatment of the residual copper and the liquated lead.

1. 100 lbs. of copper, containing about 2 ounces of silver, is melted with 250 lbs. of lead and poured into a casting pan, 2 ft. in diameter and $3\frac{1}{2}$ in. thick, forming a circular cake.

2. A number of these cakes having been cast, they are placed on the liquation hearth (Fig. 32), which consists of two walls covered with iron plates inclining towards a median line, leaving a space of 2 to 3 inches through which the liquated metal drops and runs along a gutter into a basin outside. The cakes are surrounded by an iron frame and packed round with charcoal to exclude air. A wood fire is then lighted underneath and the temperature regulated by a damper in the back wall. In about an hour the lead begins to liquate out, carrying the silver with it; when the flow ceases, the fire is allowed to die out. The residual copper contains about one third its weight of lead, with some silver; the liquated lead contains 2 to 3 per cent. of copper.

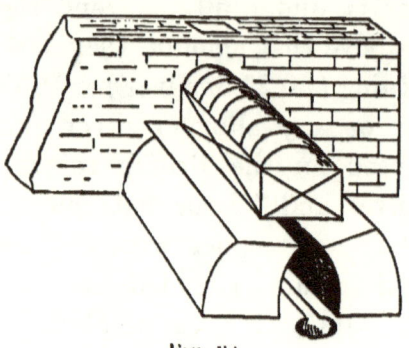

Fig. 32.

3. The residual copper containing lead and silver is exposed to a prolonged high temperature in a hearth furnace provided with a twyer, by which means the lead and some copper are oxidized, leaving an impure copper which is afterwards refined. The liquated lead is treated by cupellation for the recovery of the silver. (See Chap. VIII.)

AMALGAMATION.

1. **Patio or Mexican method.**—This process is employed for the treatment of sulphides containing silver.

The ore is first crushed by Cornish stamps worked by a water wheel, and then ground to a fine powder between two millstones called an arrastra, turned by mules.

The finely ground ore is next transferred to the amalgamating floor or patio and left to thicken by evaporation of the water; then salt earth, containing about 128 lbs. of sodium chloride (NaCl) to the ton, is added, well mixed by shovels, and afterwards thoroughly incorporated by being trodden by horses, or mixed by a mechanical contrivance. The next day copper sulphate is added, then mercury, and the operation of mixing continued until the whole of the mercury is taken up, when fresh mercury is introduced. One part of silver in the ore requires 6 parts of mercury to completely amalgamate it.

The materials in the last operation form a mud which is transferred to circular stone vats and mixed with more mercury and water. The whole is then well agitated by means of a paddle turned by mules, water constantly flowing through, by which means the mercury and silver amalgam falls to the bottom, and the earthy matter is washed away.

The excess of mercury is now removed by squeezing the amalgam in canvas bags and the amalgam taken to the "burning house." This contains an iron stand on which the compressed amalgam is placed, having a cistern below containing water, which is frequently changed. The whole is covered with an iron or copper bell heated externally by a fire; the heat vapourizes the mercury, which condenses in the water below, leaving the silver on the stand.

The changes occurring during the process may be briefly stated as follows :—The salt converts the copper sulphate into cupric chloride, which in turn is acted on by mercury forming cuprous chloride ; these two chlorides of copper decompose silver sulphide, forming silver chloride, which is decomposed by mercury, liberating metallic silver, the latter being dissolved by mercury forming silver amalgam, thus :—

$$2NaCl + CuSO_4 = CuCl_2 + Na_2SO_4$$
$$2CuCl_2 + Hg_2 = Cu_2Cl_2 + Hg_2Cl_2$$
$$2CuCl_2 + Ag_2S = 2AgCl + Cu_2Cl_2 + S$$
$$2Cu_2Cl_2 + Ag_2S + 3O = 2AgCl + CuCl_2 + 3CuO + S$$
$$2AgCl + Hg_2 = Ag_2 + Hg_2Cl_2.$$

2. **European method or Barrel amalgamation.**— This method is employed for sulphides, regulus, speise, and black copper containing silver, at Freiberg and other places.

The ore is first ground and sifted, then roasted in a reverberatory furnace with 10 per cent. of common salt. Some of the sulphur and arsenic is expelled and the sulphides oxidized, first to sulphates then to chlorides. The roasted product is ground to a fine powder and 10 to 14 cwts. put into an oaken barrel containing scrap iron, and partly filled with warm water. The barrel is then made to revolve, by which means the iron acts more perfectly on the silver and copper chlorides, reducing them to the metallic state. After revolving two hours, 3 to 5 cwts. of mercury are added and the revolution continued for eighteen to twenty hours. When the ores are very rich, scrap copper is used instead of iron.

FREIBERG PROCESS. 91

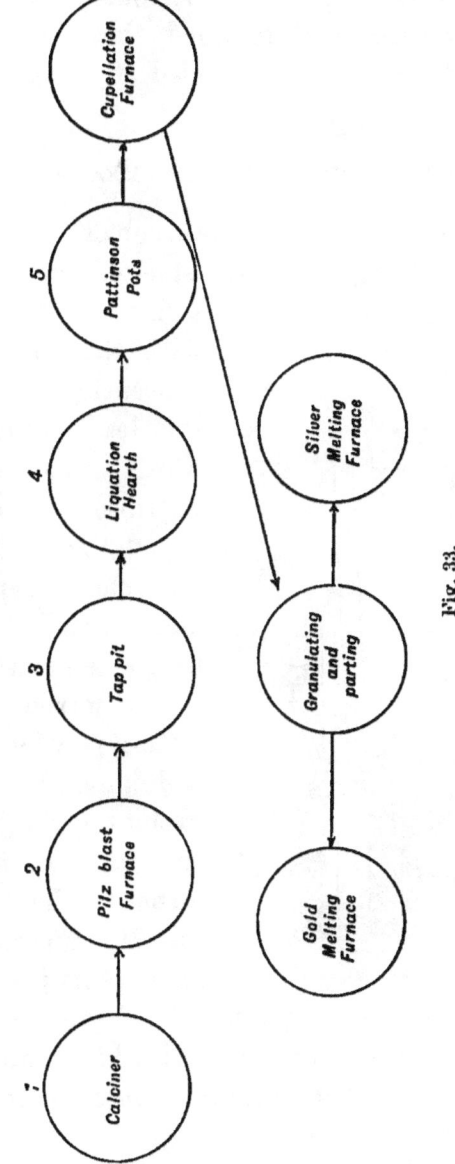

Fig. 33.

When the amalgamation is complete, the barrel is filled with water and slowly revolved for one and a half hours, the amalgam settling at the bottom. The amalgam is collected, washed, and strained through canvas, then pressed, and the mercury distilled off in an iron retort.

Extraction of Silver by means of Lead.

Ores of copper, lead, nickel, cobalt, etc., and various residues containing silver are smelted with lead ores, if sufficient is not present with the silver, so as to concentrate the silver in the lead, which is afterwards separated by the Pattinson process, and by cupellation. The foregoing scheme represents the method carried out at Freiberg (see preceding page).

The mixed ores in the state of powder are calcined and cintered to cause them to clot together, then smelted with coke in the blast furnace (Fig. 34), called the Pilz furnace, from the name of its inventor. The furnace is octagonal in plan. The hearth is surrounded with a hollow iron casing, through which water circulates to keep it cool. The throat is closed with a cup and

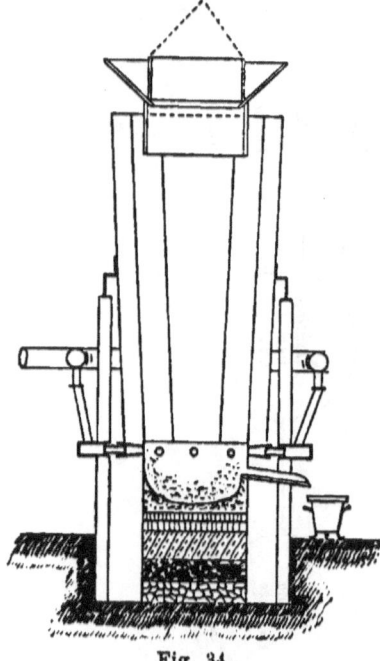

Fig. 34.

cylinder arrangement for charging and collecting the waste gases, which pass off by a side pipe and are utilized for heating the blast, boilers, etc. The operation being completed, the charge is tapped into conical moulds, and forms three layers. Argentiferous and auriferous lead and copper at the bottom, then regulus containing a little silver, and on the top of this is the slag. When nickel, arsenic, and cobalt are present, a layer of speise is formed between the lead and the regulus. The regulus, speise and part of the slag are re-smelted with another charge. The lead is treated by liquation, then by the Pattinson process, then cupelled. The gold and silver alloy is treated with sulphuric acid which dissolves out the silver and leaves the gold. The silver in solution is precipitated with a solution of common salt, as chloride, and reduced by copper or zinc to the metallic state.

Wet Methods.

Two wet methods are in common use, named after their inventors—Augustin and Ziervogel.

Augustin's method consists of roasting the ore with salt to form silver chloride, then dissolving the chloride in hot brine and precipitating the silver with copper.

The ores, chiefly sulphides and other products, are first roasted to form oxides and expel volatile matter; then roasted in a reverberatory furnace, at a low temperature, with common salt to form silver chloride.

The roasted ore is next placed in a row of tubs (Fig. 35) having false bottoms; the spaces between the false and real bottoms are filled with layers of straw and cloth through which the liquid filters, and may be

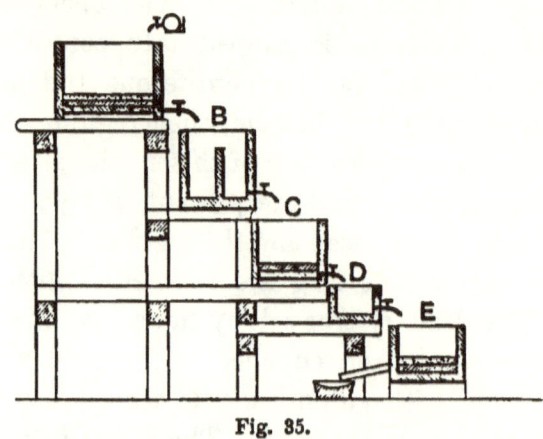

Fig. 35.

Fig. 35 a

drawn off by a tap opening into this space (Fig. 35 a), and the liquid run into a series of settling tubs placed below. About 8 cwts. of hot roasted ore are put in each tub and on to this is delivered from a tank at a higher level a hot solution of common salt, which dissolves the silver chloride, and running through the filter is tapped into settling tubs (B) for any solid matter to subside. The clear liquid is then run into the next series of tubs below (C), which have false bottoms like the first, but containing also a layer of cement copper, 6 inches thick, which precipitates the silver, as shown by the following equation—

$$Cu + 2AgCl = CuCl_2 + Ag_2.$$

The copper passes into solution and any silver remain

ing in solution is completely precipitated in the next row of tubs below (D). Lastly, the copper solution is run into a still lower series of tubs (E), when the copper in solution is recovered by precipitation with scrap iron, thus—

$$CuCl_2 + Fe = FeCl_2 + Cu.$$

The precipitated silver is collected and treated with hydrochloric acid to dissolve out any copper, and then refined.

Ziervogel's method consists of roasting finely powdered silver sulphide, in admixture with copper pyrites, to silver sulphate, dissolving the sulphate in water, and precipitating the silver with copper. The perfection of the process depends on the preliminary roasting and the amount of copper and iron sulphides present. At Mansfeld, where the method was introduced, the ore contains about 80 per cent. Cu_2S, 11 per cent. FeS_2, and 0·4 per cent. of Ag_2S. The iron and copper sulphides are decomposed, liberating the sulphur trioxide, which in turn converts the Ag_2S into silver sulphate (Ag_2SO_4).

The roasting process is performed in a double-bedded reverberatory furnace. The ore is placed on the top bed for a preliminary roasting, then raked on to the lower and hotter one to finish, with frequent stirring during the roasting. The operation requires about five and a half hours. When the roasting is complete, a portion treated with water should produce a light blue liquid, and a solution of common salt added to this liquid should give a dense precipitate of silver chloride. If the silver and copper sulphates have been overheated, they will be reduced to metallic silver and oxide of copper, when a colourless liquid will be obtained by digesting a portion with water.

The roasted mass is next put into tubs with false bottoms as in the Augustin method, in charges of 5 cwts. each, and digested with hot water at 87° C., by which means the soluble copper and silver sulphides are dissolved. The liquid is then filtered, run into settling tubs, and thence to the precipitating tubs, where the silver is precipitated with cement copper as in Augustin's process. The residues, after extracting the silver, are smelted for black copper. About 12 to 14 lbs. of silver are produced in twelve hours.

Electro-Deposition.

Silver is very extensively used for coating base metals in electroplating, the best solution for that purpose being the double cyanide of silver and potassium, which may be obtained either by the chemical or by the battery method.

1. **Chemical method.**—Dissolve 2 ounces of fine silver in dilute nitric acid; evaporate to dryness on a sand bath; re-dissolve in water, and add a strong solution of potassium cyanide, stirring all the time until the whole of the silver is precipitated; allow to settle; pour off the clear liquid and again add potassium cyanide until the precipitate just re-dissolves; now add one fourth more to form the free cyanide, and make up the solution to one gallon.

2. **Battery method.**—Make a solution of potassium cyanide containing 1½ ounces to the gallon; suspend a large silver anode and small silver cathode in the solution, then send a current through the liquid until a piece of bright and clean brass receives a good deposit.

Fig. 36 represents a silver plating vat with a copper-zinc battery attached. The silver plate A is connected with the positive (+) or copper plate of battery and is called the "anode," or the part where the current enters the liquid; the article (B) to receive the deposit is connected with the zinc or negative (−) plate of the battery and is called the "cathode," or the part where the current leaves the liquid.

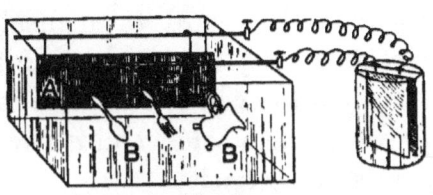

Fig. 36.

The work to receive a deposit must be scrupulously clean, for which purpose, it is generally boiled in a solution of potash to remove grease, then well swilled in clean water. In the case of copper, brass, or German silver the cleaned articles are dipped in a solution of mercury, by which the surface is coated with a film of that metal, which greatly aids the perfect deposition of the silver. Articles of zinc and iron are generally coated with a thin film of copper, in a copper cyanide solution, before immersing in the silver bath. Articles of tin, pewter and Britannia metal are taken from the potash solution to the silver solution without swilling in water, but for these metals a special silver solution is generally kept.

G

Silver may be stripped from old plated articles, without affecting the base metal, by using a mixture of 3 lbs. of sulphuric acid diluted with 1 lb. of water, and 1½ ounces of nitre.

Gold.

Gold is a yellow metal by reflected light, but green by transmitted light, with a brilliant lustre; it exceeds all others with regard to malleability and ductility; its specific gravity is 19·5; its melting point is about 1200 C.; it is almost as soft as lead, and can be welded together by pressure when cold; and is one of the best conductors of heat and electricity.

Gold does not oxidize in air, nor is it acted upon by any single acid except selenic, but is dissolved by chlorine or substances like aqua regia ($NO_3H + 2HCl$) which yield chlorine. Chloride of gold ($AuCl_3$) is a deep red crystalline powder obtained by dissolving gold in aqua regia; it is soluble in water, and is used for preparing other gold salts. The metal may be reduced from a solution of the chloride by means of oxalic and sulphurous acids, by sulphate of iron, and by most metals. Sulphur does not combine with gold at any temperature.

Alloys of gold.—Pure gold is too soft for general use, so that it is usually alloyed with silver and copper, which harden it without seriously reducing its malleability.

Fine gold is known by jewellers as 24 carat, and different alloys as so many carats fine, thus—18 carat gold contains 18 parts gold, and 6 parts copper and silver.

The gold coinage of this country corresponds to 22 carat or 916·666 parts of gold and 83·334 parts of copper per thousand.

The colour of gold is modified by the alloying metals—18 parts gold and 6 parts copper is reddish; 18 parts gold and 6 parts silver has a green tint; 18 parts gold and 6 parts iron has a blue tint; and 12 parts gold with 12 parts silver forms a white alloy.

In the case of gold solders, zinc is sometimes added to lower the fusing point.

Gold and mercury unite in the cold, but much quicker when heated, to form amalgams, which may be liquid, pasty, or solid.

Platinum is often present in gold, and does not seriously affect the malleability, but makes the colour paler.

Malleable alloys of gold and tin may be formed, if the tin is pure, producing contraction, and having a specific gravity above the mean of the constituents.

Antimony, arsenic, and lead are most injurious to gold even when present in minute traces.

Pure gold may be obtained by dissolving 1 oz. of ordinary fine gold in aqua regia, evaporating the solution to dryness, re-dissolving in water, and precipitating any platinum with potassium chloride. The solution is then diluted to a gallon, when any silver will, in the course of a few weeks, be precipitated as chloride. The clear liquid, after filtering, is then treated with sulphurous acid which precipitates the gold; this, after washing with hydrochloric acid, water, ammonia, and finally with water, may be considered pure, and contains 999·6 parts of gold per 1000. The precipitated gold is melted with hydrogen potassium sulphate and borax as fluxes.

Ores of gold.—Gold is usually found in the metallic state in nature (generally associated with silver and sometimes with copper, iron and platinum), in the state of grains in sand or alluvial deposits; in veins of quartz; and occasionally in lumps, which, when of notable size, are called nuggets. It is often found in other ores, such as galena, blende, iron and copper pyrites, etc.

EXTRACTION OF GOLD FROM ITS ORES.

Gold is extracted from its ores by means of mercury, lead, iron, or chlorine, with or without the aid of electricity.

1. **Amalgamation with mercury.**—This method is mostly used for gold quartz and finely divided ores, such as gold sand. The ore is first crushed to powder, if necessary, with stamps, and roasted if sulphides and arsenides are present. The ore is then washed forward to the amalgamating mill with a stream of water. These amalgamators vary in shape, but the Hungarian Mill (Fig. 37) will illustrate the principle on which they are based. It consists of a cast iron basin fixed on a table, and passing through its centre is the driving shaft for a wooden muller, connected with it by two iron rods, and having iron ribs on its under surface. The mercury and ore mud occupy the space between this and the bottom of the

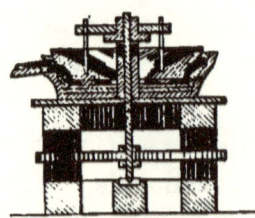
Fig. 37.

basin. The muller is hollowed out on the top side to form a basin for admission of the ore which enters the space at the centre. A 16 inch basin requires 50 lbs. of mercury.

A number of these mills are arranged in a train, so that any gold which escapes amalgamation in one may be retained by the succeeding ones. The union of the gold with mercury is greatly facilitated by the revolution of the muller, which is actuated by the central spindle. The latter carries a spur wheel which gears with a similar wheel on the next mill.

The amalgam is squeezed in bags to drain off excess of mercury, and the mercury distilled off in iron retorts, as with silver amalgam.

Mr. Rowland Jordon, of London, has introduced an automatic arrangement for the amalgamation of gold ores without the use of fire or water. The ore is first roughly crushed, then passed to a fine crusher in which the ore is subjected to the alternating action of eight flat surfaces of hard iron arranged on two spindles, each revolving with great velocity and in opposite directions. A current of air is drawn into the chamber by means of vanes on the revolving arms, by which the ore, when reduced to powder, is carried forward to a settling chamber, and there falls by its own gravity to an apparatus which automatically regulates the feeding of the powder to the amalgamator.

The amalgamator contains a layer of mercury 30 to 50 inches deep, at the bottom of which, the ore powder is delivered, and being lighter than mercury, rises to the top. The gold is dissolved by the mercury and the sand is carried away by an air current to the waste pit, or, if necessary, to another machine.

Electro amalgamation of gold ore.—In the usual method for amalgamating gold ore, in which the crushed mineral is carried over mercury by a stream of water, it sometimes happens that the surface of the mercury becomes coated with oxides, sulphides, or arsenides, which necessitates a strong force of water to remove them, and causes loss of gold. Mercury contaminated in this way is said to be "sick," and requires to be periodically distilled in order to purify it.

Mr. Barker, F.G.S., has introduced a method of preventing this "sickening" of the mercury by connecting the mercury with the negative pole of a battery or dynamo, and the water flowing over it with the positive pole, so that the current flows from the water to the mercury; by this means any oxides, sulphides, etc., are decomposed, and the black scum retreats from the surface of the mercury as if driven by a blast of air.

Electro deposition of gold.—As already mentioned, most metals and many salts deposit gold from its solutions, especially when made alkaline with carbonate of potash, potassic cyanide, etc. The deposition is more perfect when a strong electric current from a battery or dynamo is sent through the liquid.

The "Cassel" process consists of passing an electric current through a solution of common salt containing gold ore in suspension, by which the salt is decomposed, liberating chlorine; this dissolves the gold, forming chloride of gold, which is then dissolved by the liquid, and subsequently decomposed, the gold being deposited at the cathode.

The arrangement consists of a vat partly filled with a solution of common salt, and within this a barrel revolves containing the ore. This barrel is perforated with holes and covered with asbestos cloth. The interior of the barrel carries a number of carbon electrodes connected through the axis and a rubbing contact with the positive pole of a dynamo machine. The other pole of the dynamo is connected with a large copper plate placed in the outer vat, forming the cathode. By rotation of the barrel, the chlorine separated from the salt comes in contact with every particle of gold and dissolves it. To prevent the precipitation of gold in the barrel by iron salts, lime is added.

Electro gilding.—The preparation of a gold cyanide solution for gilding is very similar to that described for silver plating, but a stronger battery current is required and the solution is worked hot.

Separation of Gold from Silver and Copper.

Both wet and dry methods are employed. In the wet method either sulphuric acid or nitric acid may be used, the former being considered more economical on the large scale, but nitric acid is invariably used on the small scale, as in assaying.

Sulphuric acid method.—The alloy is made so as to contain not more than 35 per cent. of gold and 10 per cent. of copper, the rest being silver. It is then granulated by pouring into water, and placed in charges of

3 to 5 cwts. in pots made of white cast iron, heated by a fire beneath. The silver and copper are dissolved, forming sulphates, and sulphurous acid escapes, which is conducted by leaden pipes into condensing chambers. The liquors are then siphoned off and the silver precipitated by copper. The gold left behind still retains silver, and is boiled with fresh acid (sometimes in platinum vessels) to remove the last portions. The gold is then washed, dried and melted in plumbago crucibles with a little nitre and borax. The copper sulphate solution is concentrated and crystallized.

Nitric acid method.—The prepared alloy is boiled first with diluted acid, then with strong acid, in glass or platinum digesters. The liquors containing the silver are treated with copper or zinc to precipitate the silver, and the gold washed and melted as in the above method.

Dry method by means of chlorine.—The gold alloy is melted in a clay crucible fitted with a lid, having a hole in the centre through which a clay pipe passes to the bottom of the metal for conveying the chlorine gas. The metal being melted, the surface is covered with borax, and the chlorine passed until orange coloured vapours appear, showing that all base metals and the silver have been chlorinized, and that the gold is beginning to be attacked. The contents are then allowed to cool, when the gold solidifies and the liquid chloride of silver may be poured off. This process is valuable when the amount of silver is small, but when much silver is present, the wet methods of parting are preferable.

Platinum.

Platinum is a white metal, with a brilliant lustre; highly malleable and ductile; as soft as silver, and can readily be welded; it is very tenacious, being only exceeded by iron and copper among the elementary metals; it only melts at the highest temperatures, such as those of the oxy-hydrogen flame and the electric arc. It does not oxidize at any temperature, and resists the action of all single acids, its best solvent being aqua regia. It is one of the heaviest metals, having a specific gravity of 21·5. Like silver it absorbs oxygen when melted, giving it out again on cooling, causing the mass to spit. It absorbs considerable quantities of hydrogen and other gases when heated, especially the spongy variety called platinum black; if this substance be introduced into a mixture of oxygen and hydrogen it causes them to combine with the development of great heat.

Platinum occurs in nature like gold, in the metallic state, in the form of grains or nuggets, often associated with iron, copper, gold, silver, and several rare metals.

Methods of Extraction.

1. **Wet method.**—The ore is first washed and gold and silver separated by amalgamation. 2. The residue is next digested with boiling nitric acid to dissolve out all base metals, such as iron and copper. 3. It is then boiled with aqua regia, forming a solution of platinum chloride. 4. The solution is then evaporated to remove the acid; re-dissolved in water; mixed with an equal

bulk of alcohol, and a solution of ammonium chloride added, which precipitates the platinum as a double chloride ($2NH_4Cl, PtCl_4$). 5. This chloride is washed, dried, and ignited in a plumbago crucible, when the platinum is left as a black powder. 6. The spongy mass is then welded by heat and pressure into the compact form.

Dry method.—Deville and Debray extract platinum by smelting the ore in a reverberatory furnace with galena or litharge, thus forming an alloy of lead and platinum, which is afterwards cupelled. The cupellation cannot be finished in an ordinary furnace in consequence of the high melting point of platinum, so that the refining is completed on the lime hearth of the oxy-hydrogen furnace (Fig. 38). This consists of two hollow blocks of lime placed together and bound with iron. The upper block is perforated for the introduction of the blowpipes, one tube of each pair conveying coal gas and the other oxygen. The metal is placed in the cavity of the lower block, and 20 to 30 lbs. may be refined in an hour.

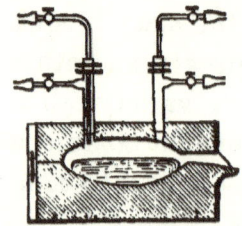

Fig. 38.

The spongy platinum obtained by the wet method is now melted and refined in this furnace, while, for the purpose of casting platinum into ingots, a crucible of gas coke is used in which the metal is placed; this is enclosed in a lime crucible, and the whole heated in a lime furnace by means of the oxy-hydrogen blowpipe flame.

Electro deposition.—There are two names given to the electro deposition of platinum—platinizing and platinating. By the former is meant its deposition as a dark film, and by the latter its deposition as a white metal. Smee platinized sheets of silver for his battery, and various shades of colour are imparted to different parts of silver-plated goods by painting with solutions of platinum chloride of different strengths.

A plating solution may be made by dissolving platinum chloride in potassium cyanide. The solution is used warm, with a moderately weak current. The anode does not dissolve, so that the solution must be periodically replenished with fresh platinum salt.

CHAPTER VII.

Copper (Cu).

Copper has a red colour; is highly malleable, ductile and tenacious; melts at about 1100° C.; is not sensibly volatile except at very high temperatures; its specific gravity is 8·96, which may be slightly increased by hammering and rolling. Copper is one of the best conductors of heat and electricity, but this property is considerably reduced by the presence of small traces of foreign substances. It oxidizes at a red heat, and is soluble in acids, forming copper salts. The most common impurities in the commercial metal are iron, arsenic, silver and oxide of copper; occasionally bismuth, tin, antimony, sulphur and lead; and of these, arsenic and antimony are the most injurious.

Copper forms two compounds with oxygen—black or cupric oxide (CuO) and red or cuprous oxide (Cu_2O). When copper is heated in air the red oxide is formed, coated on the outside with a thin scale of black oxide. When either of these oxides is heated with silica, a silicate of cuprous oxide is formed thus—

$$2CuO + SiO_2 = Cu_2O,SiO_2 + O$$
$$Cu_2O + SiO_2 = Cu_2O,SiO_2.$$

Cuprous oxide is soluble in molten copper, forming

dry or underpoled copper. When copper is exposed to air and moisture, it first oxidizes, then forms a green carbonate by the action of carbonic acid.

Copper unites directly with sulphur when the two elements are heated together, forming cuprous sulphide (Cu_2S). Also, when sulphide of iron and red oxide of copper are heated together a mutual exchange takes place, forming cuprous sulphide and oxide of iron, which, in presence of silica, forms silicate of iron. Cuprous sulphide is oxidized when heated in air or oxygen, and at a high temperature, with steam, thus—

$$Cu_2S + 3O = Cu_2O + SO_2$$
$$Cu_2S + OH_2 = Cu_2O + SH_2.$$

When cuprous sulphide is heated with cuprous or cupric oxide or sulphate in suitable proportions, metallic copper is isolated thus—$Cu_2S + 2Cu_2O = SO_2 + 6Cu$. When cuprous sulphide is heated with iron, a double sulphide of iron and copper is formed, and some copper set free; it is also partially decomposed by carbon.

Copper and Carbon do not unite, but when commercial copper is heated with carbon, the oxygen it contains is removed and overpoled copper produced, which is red short.

Copper and Phosphorus.—These elements readily unite at a red heat, forming a hard, nearly white, brittle phosphide. Even in small quantities, phosphorus affects the physical properties of copper. Thus $\frac{1}{2}$ per cent. makes copper harder, more tenacious, and more fusible.

ORES OF COPPER.

Native copper often occurs with other ores, and is sometimes covered with a crust of oxide and carbonate. The richest deposits of native copper occur on the shores of Lake Superior. Sometimes it occurs as grains in sand.

Red oxide or *cuprite* (Cu_2O) may contain 88·78 per cent. of copper and the *black oxide* (CuO) 79·82 per cent. *Green carbonate* or *malachite* ($CuCO_3 + CuH_2O_2$) may contain 57 per cent. of copper and possesses a fine emerald green colour. *Blue carbonate* or *azurite* ($2CuCO_3 + CuH_2O_2$) may contain 55·16 per cent. of copper. *Copper pyrites*, or *yellow copper ore* (Cu_2S, Fe_2S_3), is the most abundant British ore, and contains when pure 34·81 per cent. of copper. It may be distinguished from iron pyrites by its superior softness. *Vitreous*, or *grey sulphide of copper, copper-glance, redruthite* (Cu_2S) may contain 79·7 per cent. of copper. *Fahlore*, or *grey copper ore*, is a sulphide of copper where the copper is partly replaced by other metals, and often contains silver. *Chrysocolla*, a hydrated silicate of copper, and *atacamite*, a hydrated oxy-chloride, are sometimes found.

METHODS OF EXTRACTION.

There are four main processes of extracting copper—1. Reverberatory method; 2. Blast furnace method; 3. Combination of 1 and 2; 4. Wet and electro-chemical methods.

Welsh method.—This may be explained by reference to the following scheme :—

EXTRACTION OF COPPER.

SCHEME OF THE WELSH METHOD.

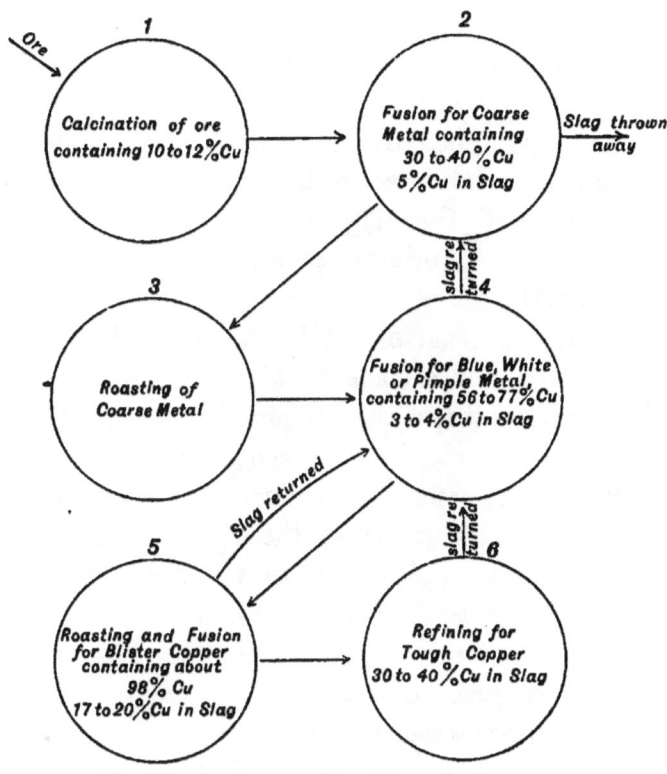

Fig. 39.

1. Calcination with air.—Volatile matter, such as arsenic, water, carbonic acid, sulphurous acid, etc., is expelled; some iron is oxidized, and the product contains iron copper, silicon, sulphur, oxygen, and earthy matter.

2. Fusion for coarse metal.—The charge consists of calcined ore, metal slag, and a little fluor-spar. The products are a regulus of iron and copper, and an acid silicate, chiefly of iron ($FeOSiO_2$). If insufficient sulphur is present, some copper is reduced and the slag becomes too rich. Coarse metal is a bronze-coloured, stony-looking porous mass, and the slag is a dark-coloured brittle substance, often containing pieces of quartz and bits of coarse metal.

3. Roasting of coarse metal.—More sulphur escapes, iron is oxidized, and a black friable mass results.

4. Fusion for blue, white, or pimple metal.—The product of No. 3 is melted with roaster and refinery slags, and a certain proportion of oxidized ores, such as oxides and carbonates, which re-act on the regulus, the object being to convert the whole of the iron sulphide into oxide, leaving the copper as sulphide. When the oxidized ore is insufficient, blue metal results, which contains sulphide of iron; when just sufficient, white metal, which is sulphide of copper, results; when in excess, some of the sulphide of copper is reduced, and the surface of the metal has a number of pimples due to the escape of sulphurous acid. It is then termed pimple metal.

Blue metal is a brittle substance with an uneven fracture, of a purplish blue colour when broken hot, and a bronzy tint when broken cold. It is often covered with fine strings of metallic copper called "moss" copper.

White metal is compact, brittle, with an uneven granular or crystalline fracture and of a bluish grey colour.

The slag is compact, brittle, with an uneven fracture, leaving sharp edges; granular and sometimes crystalline, of a dark bluish grey colour, with bubbles on the surface. It is termed metal slag.

5. Roasting and fusion for blister copper.—The products of No. 4 are roasted so as to oxidize a portion of the copper sulphide, then melted, when the sulphide and oxide re-act, as shown in the following equation—

$$Cu_2S + 2Cu_2O = SO_2 + 6Cu.$$

The surface of the ingot is covered with blisters which are due to the escape of SO_2 after the metal has solidified. The slag is sometimes light, porous, and of a black colour, sometimes dense and close, containing shots of copper, and of a reddish brown or black colour.

6. Refining blister copper.—The copper is quickly melted, and the metallic oxides present, together with much oxide of copper, combine with the silica to form a fusible slag. Such bodies as sulphur, antimony and arsenic are volatilized. The copper still retains some impurities and red oxide of copper. These are removed by covering the metal with anthracite powder and inserting a pole of green wood which produces reducing gases, such as CO and CH_4, which act chemically, in reducing oxide of copper, and mechanically, by bringing the impurities to the surface. If the poling is insufficient the copper is

dry in appearance, dull in lustre, and cracks at the edges when hammered, called "underpoled" or dry copper. When the right degree of poling is effected it is said to be at tough pitch. If the poling be carried too far, the copper loses its malleability and ductility. This is probably due to the reduction of the oxides of foreign metals, such as lead and antimony, which alloy with the copper. Overpoling may be remedied by admitting a little air to the copper. An ingot of overpoled copper has a longitudinal ridge on its surface, and underpoled copper has a longitudinal depression, while tough copper has neither ridge nor furrow, but a level surface.

Calciners.—The furnaces employed for calcining copper ores have a long bed and comparatively small fireplace; the roof is low and contains two or more hoppers for charging, according to the size. An old form for calcining 3 or 4 tons is shown in Fig. 40. It

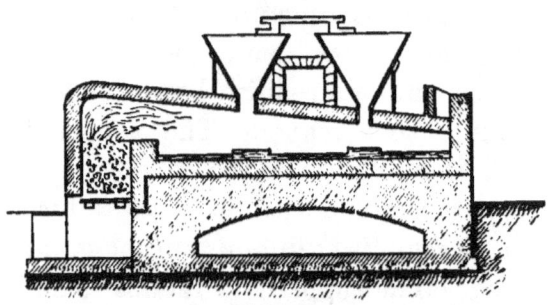

Fig. 40.

is lined with fire-brick, the bed being about 16 ft. long and 12 ft. wide. Several attempts have been made to utilize the sulphur given off during calcination. Mr. Spence invented a furnace for this purpose, 50 ft. long, heated from beneath, so that the furnace contains two chambers—an upper one in which the ore is placed and a lower one through which the products of combustion pass. The sulphur liberated from the ore in the form of SO_2 is utilized for the manufacture of sulphuric acid. The Gerstenhöfer calciner (Fig. 41) is another arrangement for the same purpose. It is a

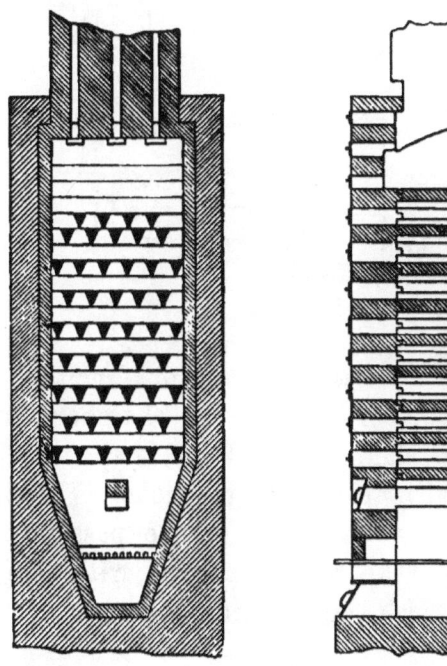

Fig. 41.   Fig. 41 a.

rectangular chamber 20 feet high and 5 feet across, in which are arranged 20 rows of triangular fire-clay bars, 2 ft. 6 in. long. The ore is ground fine, and the supply, regulated by feed rollers, passes through three narrow channels, and falls on the flat surface of the bars, then gradually descends as fresh ore is added, being oxidized in its descent, and is finally removed at the bottom. Air is admitted by a series of apertures in the front of the furnace and the combustion of the sulphur maintains the necessary temperature.

Melting furnace.—The melting furnace has a smaller bed and larger fireplace than the calciner.

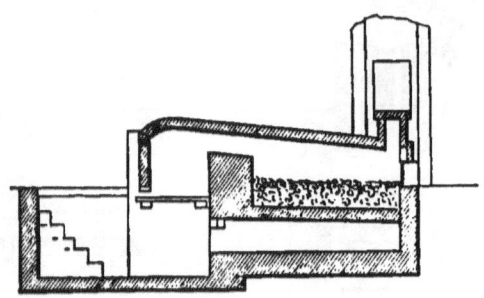

Fig. 42.

Fig. 42 shows a section of an ore melting furnace. The bed or laboratory is an oval chamber, lined with sand and metal slag, 13 ft. × 9 ft., shaped hollow, and inclining from all parts to the tap hole. The fire-bridge is generally hollow to allow for the passage of air to keep it cool, and the roof is pierced with one opening communicating with the charging hopper.

BLAST FURNACE METHOD.

This plan is adopted where fuel is dear, the ores poor, refractory, and not too complex. The following method carried out in Sweden may be taken as a type.

1. The ore is roasted in heaps, 30 ft. square and 12 ft. high, or in stalls composed of three walls; the burning

of the sulphur, together with some wood, maintains the heat.

2. The calcined ore is then melted in the blast furnace (Fig. 43) with slags obtained from the black copper in No. 4. The furnace is 18 ft. high 3 ft. 10 in. wide at the twyers, and the hearth 5 ft. 4 in. deep. In from two to three days the hearth is filled with regulus containing about 30 per cent. of copper, which is tapped into sand moulds.

3. Roasting the regulus from No. 2 called raw "matt." This is conducted in heaps, 11 ft. × 5 ft. × 5 ft. high, each heap containing about 5 tons of matt, and requiring seven to eight weeks for completion.

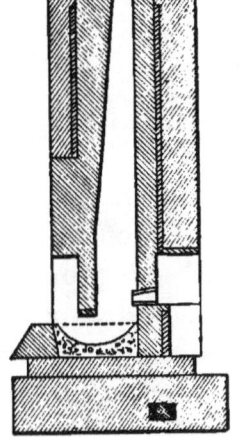

Fig. 43.

4. Smelting for black copper.—The roasted regulus is rich in metallic oxides and requires a siliceous flux, such as ore slag. The charge consists of 2 cwts. calcined regulus, 28 lbs. of copper residues, 28 lbs. of ore slag, 12 lbs. of quartz and 18 lbs. of charcoal. One operation lasts from two to three days, and the products are—black copper containing about 94 per cent. of copper, a regulus containing 60 to 70 per cent. copper, and a slag composed chiefly of silicate of iron (FeO, SiO_2). The furnace is represented in Fig. 44.

Fig. 44.

5. Refining black copper.—The refining is conducted

in the blast hearth (Fig. 45), 2½ ft. × 2 ft. diameter, and 15 to 18 in. deep, lined with fire-clay and sand. The twyer inclines at an angle of 45°. The copper is melted with charcoal by the aid of a blast of air, and the bath of metal kept covered with charcoal during refining. The slag is removed two or three times and when it becomes red from the presence of oxide of copper the blast is turned off, the charcoal pushed back and the slag solidified by projecting water on to the surface, and then removed as a cake. The copper is kept covered with charcoal until sufficiently refined or until the excess of cuprous oxide has been reduced by the carbon.

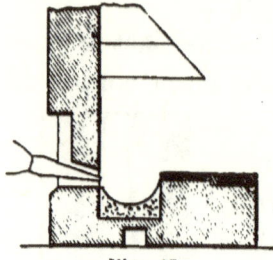

Fig. 45.

Kernel roasting. — This is an ancient process, formerly conducted in many parts of Europe and now carried out at Agordo in the Venetian Alps, the ore having an average composition of 2 per cent. of copper. It is a cupriferous iron pyrites and when roasted in lumps, 2 to 3 inches square, the copper becomes concentrated in the centre, forming a nucleus consisting essentially of copper, iron, and sulphur, containing 4·5 per cent. of copper. The outer shell consists chiefly of oxide of iron, with a little copper oxide and some copper sulphate, which latter is dissolved out with water.

The calcination was conducted formerly in heaps but is now performed in kilns. The kiln (Fig. 46) is a rectangular chamber within four walls divided into sections. The bed of each section is made in the form of a

EXTRACTION OF COPPER.

pyramid, on the top of which a chimney is roughly built with pieces of ore, and the spaces between are filled with

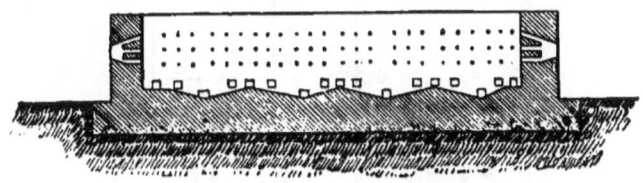

Fig. 46.

the ore to be roasted. The kernels obtained are smelted by the blast furnace method to obtain metallic copper.

WET METHODS.

In some cases copper pyrites, by the action of air and moisture, is converted into copper sulphate, which is dissolved by water, forming a solution from which the copper is precipitated by iron, thus—

$$CuSO_4 + Fe = FeSO_4 + Cu.$$

One of the most successful methods is due to Henderson and largely carried out by the Tharsis Company. The ore is first roasted to remove the greater part of the sulphur for the manufacture of sulphuric acid, leaving the iron as oxide, and the sulphur in about equal amount to that of the copper present. The ore contains 2 to 3 per cent. copper and small quantities of silver and gold, which are recovered by the Claudet process. The mode of procedure may be explained by means of the following scheme (Fig. 47).

1. The roasted ore is mixed with salt, ground, sifted

120 ELEMENTARY METALLURGY.

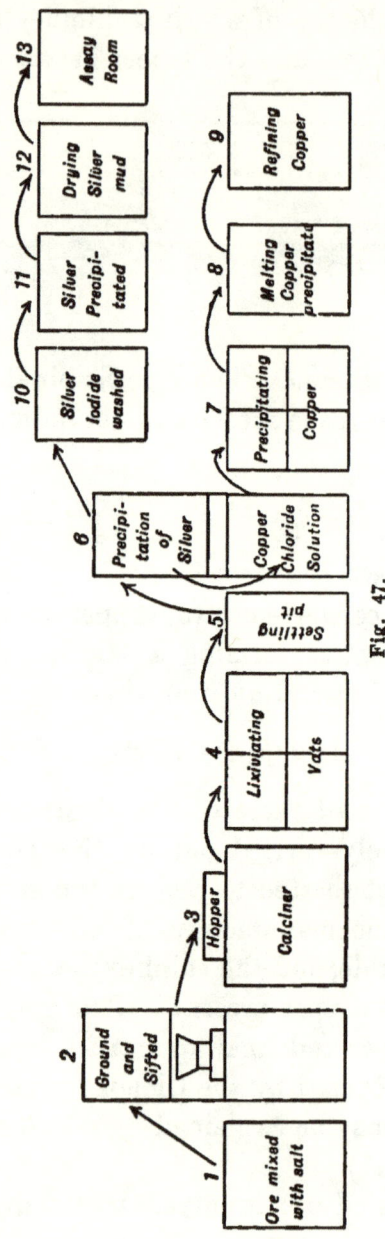

Fig. 47.

and roasted, forming copper chloride and sodium sulphate, which are dissolved in acid liquors.

$$CuSO_4 + 2NaCl = CuCl_2 + Na_2SO_4.$$

2. The silver in solution is then precipitated by a solution of zinc iodide, and the copper liquor siphoned off, leaving the silver iodide as a powder at the bottom.

3. The copper chloride solution is run into vats, and the copper precipitated by means of scrap iron, $CuCl_2 + Fe = FeCl_2 + Cu$. The precipitate is washed, then melted and refined in reverberatory furnaces, as in the Welsh method.

4. The silver iodide is washed and the silver displaced by means of zinc, whereby zinc iodide is reproduced. The chemical changes are shown by the following equations:—

$$2AgCl + ZnI_2 = 2AgI + ZnCl_2$$
$$2AgI + Zn = ZnI_2 + Ag_2.$$

Gold, if present, goes with the silver.

Electrolytic Extraction of Copper.

In depositing copper by means of electricity it is necessary that the metal should be in solution, copper sulphate being generally employed. If copper regulus is to be treated, it is cast into flat slabs, rods of copper being placed in for supports and connections; these slabs form the anodes in solutions of copper sulphate. The current is obtained from a dynamo machine.

The impurities, especially arsenic, antimony, and bismuth, introduce many difficulties in the way of successful working, and great experience is required to overcome them. The principle of the process rests on the fact that when the anode or solution contains several metals, the current makes a selection, and deposits them in a definite order.

That metal will be first dissolved from the anode the solution of which causes the development of the greatest amount of energy, and that metal will be first deposited from solution the separation of which requires the least consumption of energy. Thus, suppose the following metals present in the anode—Zinc, iron, tin, nickel, lead, arsenic, bismuth, antimony, copper, silver, and gold. All those which precede copper will be dissolved first, and those which follow copper will be dissolved last. When the above metals are in solution, the gold will be the first to be deposited and the zinc last. But this rule is dependent on the strength of current, nature of solution, whether acid or alkaline, and the proportions of the constituent metals. If the current exceeds a certain strength all may be dissolved and deposited together, and the more neutral the solution the more easily will the electro negative metals be dissolved, and the more easily will the electro positive metals be deposited. If a copper anode contains a large amount of impurities these will be more readily dissolved than from copper containing only a small amount. The less dense and compact the anode is, the better will the process go on.

The foregoing remarks apply to copper containing metals in the metallic form. If they are present as oxides or sulphides, the former, being mostly non-conductors, simply fall to the bottom of the vessel, the latter, being conductors, may be dissolved and the metal deposited; but if there is much copper in the anode and little sulphide, the latter falls to the bottom and the copper only is deposited.

Electro deposited copper, from a copper chloride or sulphate solution, is remarkably pure, for silver and gold, when present, are precipitated from the solution; lead forms an insoluble sulphate; iron, zinc, and tin are too electro positive to be deposited at the same time as copper, and the same applies to nickel, although in a less degree. The greatest danger is with antimony, when that metal is present in solution with the copper.

Copper plating.—For coating articles of brass and German silver with copper, a solution of copper sulphate is employed; but articles of zinc, iron, tin, and lead require a copper cyanide solution.

Zinc (Zn).

Zinc is known in commerce under two names—in the cast state as "spelter," and when rolled into sheets, etc., as zinc. It is a white, highly crystalline metal, with a bluish shade and bright metallic lustre; when pure, it is malleable at the ordinary temperature, while commercial cast zinc is brittle; it becomes malleable and ductile however, if heated to a temperature of 100 to 150° C., but beyond that point it again becomes brittle. The specific gravity in the cast state is 6·86, which may be increased to 7·21 by rolling or forging; it contracts but slightly on cooling from the liquid state and thus forms good castings; it melts at 430° C., and boils at 1040° C. At a red heat in air it rapidly oxidizes and burns with a greenish white flame forming zinc oxide (ZnO); raised to a bright red heat in a closed vessel, it may be readily distilled. When rolled zinc is exposed to air and moisture a grey film of suboxide is formed, which preserves the metal from further oxidation. Ordinary zinc readily dissolves in dilute hydrochloric and sulphuric acids, while the pure metal is unaffected; both kinds dissolve in nitric acid and in alkalies. Zinc displaces silver, gold, platinum, bismuth, antimony, tin, mercury, and lead from their solutions. The chief impurities of the commercial metal are iron, lead, and arsenic.

Zinc and sulphur do not readily unite, but when a mixture of finely divided zinc and sulphur is projected into a red-hot crucible, some zinc sulphide is formed. It is also formed by heating zinc with cinnabar.

Zinc forms compounds with phosphorus and arsenic when these bodies are heated together, having a metallic lustre and somewhat vitreous fracture.

The chief ores of zinc are:—The oxide (ZnO) called zincite or red oxide of zinc, which is white when pure, but generally red from the presence of oxide of manganese; the sulphide (ZnS) known as "blende" and "black jack," which is the principal source of the metal, and generally black or yellowish black in colour, but sometimes it has a reddish tint from the presence of galena; when pure it is white and contains 67·03 per cent. of zinc; the carbonate ($ZnCO_3$), called calamine; and the silicate ($2ZnO, SiO_2, OH_2$), called electric calamine.

Modes of Extracting Zinc.

In consequence of the volatile character of zinc it is always extracted by distillation at about 1000° C. Blende being the chief ore is most commonly employed, and is first converted into oxide, then reduced by carbon and carbonic oxide, the latter probably playing the most conspicuous part. It may also be reduced by hydrogen and hydrocarbons, and as all these agents are present in the products of decomposition of the coal employed, each helps in effecting the isolation of the metal. Sulphide of zinc may also be reduced when heated in suitable proportion with oxide of zinc, thus—$ZnS + 2ZnO = SO_2 + 3Zn$. It is also reduced by iron.

The roasting of the powdered blende is performed in a reverberatory furnace until the ore is "sweet," that is, until no more sulphur can be driven off, and the zinc is almost entirely in the form of oxide.

The distillation is conducted in closed fire-clay retorts connected with a cool receptacle, in which the zinc vapours are condensed. The re-actions occurring in the above operations may be represented thus :—

$$ZnS + 3O = SO_2 + ZnO.$$
$$ZnO + C = CO + Zn.$$
$$ZnO + CO = CO_2 + Zn.$$

In the last equation we see that CO_2 is formed, which, being an oxidizing agent, will oxidize metallic zinc, thus acting in opposition to CO, so that a large excess of CO must be present to counteract this neutralizing agency When the two oxides of carbon are present in the proportion of 28 : 44 no reduction can occur, as shown in the following equations :—

$$ZnO + CO = Zn + CO_2.$$
$$Zn + CO_2 = ZnO + CO.$$

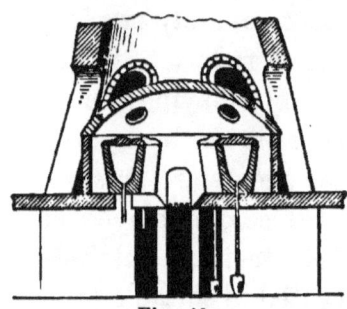

Fig. 48.

The furnaces employed for distilling zinc are of three kinds, viz., the Old English, the Belgian, and the Silesian.

The Old English furnace (Fig. 48) is a furnace of the "gallery" type, in which a row of fire-clay crucibles is placed on each side of a central fireplace. Each pot is 4 feet high, 2 feet 6 inches wide at top, closed with a lid, and perforated with a hole at the bottom, into which fits a condensing pipe for conveying the distilled metal into an iron receiver at the bottom.

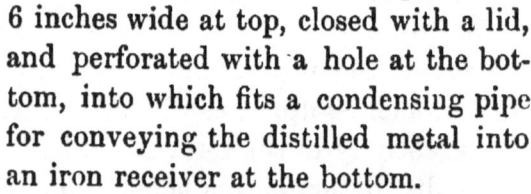

Fig. 49.

The Belgian furnace (Fig. 49) contains a vertical arched chamber in which are placed several rows of cylindrical fire-clay retorts (*a*) 3 feet 6 inches long and 8 inches in diameter, outside measure, the fireplace being below. Each retort is fitted with a simple bellied fire-clay nozzle (*b*) 12 inches long, in which the zinc for the most part condenses, and to the front of this is attached a sheet-iron tube (*c*) for condensing the "fume," which is a mixture of zinc oxide and finely divided zinc. The calcined blende is ground to powder, mixed with half its weight of small coal or coke, and about 25 lbs. of the mixture introduced into each retort, when the condensing tubes are luted on. Each furnace contains about sixty retorts, treating about 1 ton of ore in twenty-four hours, and yielding 30 to 40 per cent. of zinc. The crude zinc obtained from every mode of distillation requires to be re-melted, when zinc oxide and other impurities rise to the surface as dross. The metal is then cast into ingots weighing 70 to 80 lbs. each.

The Silesian furnace (Fig. 50) is a "gallery" furnace, containing two rows of ⌂-shaped muffles closed at one end, each 3 feet long and 8 inches wide, made of fire-clay, with the fireplace between. The upper part of the outer end of each muffle opens into an elbow-shaped clay nozzle, which connects with an iron pipe leading into a condensing chamber. The charge for each muffle is about 100 lbs.

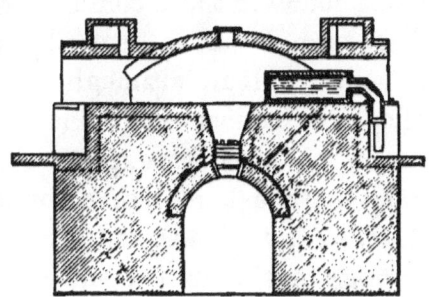

Fig. 50.

Instead of the condensing arrangement just described the Belgian system of condensation is now generally adopted. Zinc muffles and retorts are now largely heated with gas using hot air, heated either on Siemens', Boetius', or Ponsard's principle. (See Figs. 7, 8 and 26.)

Electrolytic extraction.—Létrance recommends

an electrolytic method for the deposition of zinc. The blende is roasted at a low temperature, by which means the sulphide is largely converted into sulphate, thus—

$$ZnS + 4O = ZnSO_4.$$

The product is dissolved in dilute sulphuric acid, placed in vats with a plate of carbon as the anode, and a plate of zinc as the cathode, the current being supplied by a dynamo machine.

Alloys of zinc.—The most important alloys of zinc are those with copper called "brass." These metals mix together in all proportions forming alloys of various shades of colour between red and white, the purest yellow being found in those containing 75 to 80 per cent. copper. The term brass is generally applied to those having a yellow colour, although it is sometimes used for all alloys of copper and zinc. In many cases different names are used to express the same substance. Ordinary brass is harder than copper; highly malleable and ductile; is well adapted for castings; melts at a lower temperature than copper, and is capable of taking a high polish. Occasionally small quantities of other metals are added. The following is the composition of various zinc alloys:—Tombec, 84 Cu and 16 Zn; Prince's metal, 83 Cu and 17 Zn; gilding metal, 80 Cu, 18 Zn and 2 Sn; Bath metal, 78 Cu and 22 Zn; Mosaic gold, pinchbeck, etc., 75 Cu and 25 Zn; brass, $66\frac{1}{2}$ Cu and $33\frac{1}{2}$ Zn; brass for turning, $66\frac{1}{2}$ Cu, 32 Zn and $1\frac{1}{2}$ Pb; white brass, $33\frac{1}{2}$ Cu and $66\frac{1}{2}$ Zn. Aich's, sterro, Gedge's and delta metal contain iron in addition to copper and zinc.

The formation of alloys of copper and zinc, according to Riche, is attended with contraction, which is a maximum in the alloys Zn_3Cu_2 and Zn_2Cu. These alloys are brittle and do not exhibit any of the physical properties of the constituent metals. The density of brass is increased by mechanical treatment, but this is partly annulled by heating and slow cooling. Pinchbeck is not altered in density either by working, or by slow or rapid cooling after heating.

Brass soon tarnishes when exposed to the air, and by continual vibration or stress a crystalline structure is induced which makes it brittle.

Brass is prepared by melting the copper in a black-lead crucible, or on the bed of a reverberatory furnace, and then cautiously adding the necessary quantity of zinc. In practice a certain amount of scrap brass to the extent of about one third the charge is added. A little lead is added to brass required for turning and rolling.

Galvanized iron is iron plate and iron articles coated with zinc to prevent the iron from rusting. The zinc was originally deposited by means of an electric current. It is now performed by dipping the clean iron into a bath of molten zinc, the surface of which is covered with sal-ammoniac in order to prevent oxidation, and at the same time to dissolve any oxide that may be formed. After a bath of zinc has been worked some time, an alloy of iron and zinc is formed which is duller than zinc, with a somewhat scaly appearance on its fractured surface, which is also covered with a number of grey or black specks. The zinc may be largely recovered from this alloy by distillation.

CHAPTER VIII.

Lead (Pb).

Lead has a bluish grey colour and considerable lustre when freshly cut; it is malleable, ductile, and tough, but has feeble tenacity. Pure lead emits a dull sound when struck, but the presence of impurities renders it more sonorous; it melts at 330° C. and contracts on cooling; its specific gravity is 11·44. It is so soft that it can be cut into slices with a knife, or squirted into the form of tubes or rods, and two pieces can be welded together by pressure in the cold. Sheets of lead and tin can be rolled into one compact sheet, called "autogenous soldering." It oxidizes in moist air, and is insoluble in ordinary sulphuric acid, so that lead chambers are employed in the manufacture of that acid. It is volatile when heated in air, forming lead oxide (PbO).

Oxide of lead or plumbic oxide (PbO) is yellow and exists both in the amorphous and crystalline states. The former, called massicot, is formed when molten lead is exposed to air or oxygen below a red heat, the latter, called litharge, is of a deeper colour and formed when the oxide has been fused, which occurs at a red heat. PbO is reduced by carbon, hydrogen, and the ordinary reducing agents; it also unites with many oxides and renders them readily fusible, but different oxides require different amounts to effect this fusion. Cuprous oxide (Cu_2O) requires 1 part; ferric oxide (Fe_2O_3) 4 parts; zinc oxide (ZnO) 8 parts; and tin oxide (SnO_2) 12 parts. It acts as an oxidizing agent on many metals, such as copper, zinc, iron, etc., itself being reduced.

When massicot is heated to about 300° C. it takes up oxygen forming red lead (Pb_3O_4), thus—

$$3PbO + O = Pb_3O_4.$$

The oxides of lead are reduced by hydrogen, carbon, and carbonic oxide.

Lead and sulphur unite when heated together to form lead sulphide (PbS), which is a bluish grey, brittle, and crystalline body, melting at a strong red heat. When roasted in air at a low temperature it is oxidized to sulphate ($PbSO_4$), but at a higher temperature it is converted into oxide (PbO), thus—

$$PbS + 4O = PbSO_4$$
$$PbS + 3O = PbO + SO_2.$$

If lumps of lead sulphide about the size of peas be roasted in air, the outer layers will be oxidized and an inner nucleus of unaltered sulphide will remain. If such a mass be fused, the oxide, sulphate and sulphide re-act on each other, liberating metallic lead, thus—

$$PbS + PbSO_4 = Pb_2 + 2SO_2$$
$$PbS + 2PbO = 3Pb + SO_2.$$

Sulphide and sulphate of lead are most readily reduced

by fusion with iron, forming sulphide of iron (which floats as a regulus on the top), and metallic lead, thus—

$$PbS + Fe = FeS + Pb.$$

The same effect is produced by oxide of iron and carbon instead of iron. When heated with carbon or carbonate of soda alone, only a partial reduction of the lead sulphide takes place.

Alloys of lead.—For type metal and Britannia metal see "Antimony," and for fusible alloy see "Bismuth." Shot metal is lead containing 40 lbs. of arsenic per ton, which causes the small fragments of molten metal to assume the globular form in falling from a height. Pewter contains about 80 parts tin and 20 of lead; but other metals are often added in small quantity, such as copper, antimony and zinc. Soft solder contains varying proportions of lead and tin, the best quality being that which contains most tin. Lead and zinc may be melted together in all proportions, but they separate to a great extent on cooling; if silver be present, the zinc which solidifies as a crust on the surface will contain the silver and a little of the lead. This is the basis of Parkes' process for desilverizing lead. Silver and lead alloy together in all proportions, but a certain amount of liquation takes place on cooling, whereby the outer portions of the alloy are much poorer in silver than the inner core; also if such an alloy be allowed to cool slowly and constantly stirred, the rich portion remains liquid after the other has solidified. This is the basis of the Pattinson process for desilverizing lead. Lead readily dissolves gold when melted with it, and is

EXTRACTION OF SILVER FROM LEAD.

used for the extraction of both gold and silver from their ores and products.

Ores of lead.—Galena (PbS) is by far the most abundant ore of lead and is the chief source of the metal. It is a lead-grey mineral with a metallic lustre and gives a lead-grey streak; it is generally found crystallized in cubes and has a specific gravity of about 7·6. It always contains some silver. Cerusite (PbCO$_3$) occurs sometimes in needle-shaped crystals, and sometimes massive. It is white when pure, but often has a dirty white or reddish colour and gives a white streak. Pyromorphite (3Pb$_3$P$_2$O$_8$+PbCl$_2$) is generally green in colour, but sometimes yellow or brown, and gives a white streak. Mimetesite resembles pyromorphite but contains arsenic instead of phosphorus. Anglesite (PbSO$_4$) is generally white or grey in colour.

Extraction of silver from lead.—"Pattinson's process," the principle of which has already been mentioned, is conducted in a series of 9 to 15 pots (Fig. 51), each

Fig. 51.

capable of holding from 3 to 7 tons of lead, and sometimes 15 tons, each heated by a separate fire. Suppose 3 tons of lead, containing 10 ounces of silver per ton, are first melted in the middle pot of the series, say No. 8, it is then well stirred, skimmed, and allowed to cool. At a certain temperature the poor lead separates in crystals which are removed as they form by a perforated ladle and transferred to the next pot on the right, No. 9, leaving a rich liquid argentiferous lead in the pot, which, when the operation has been carried sufficiently far, is conveyed to the next pot on the left, No. 7. This will contain 1 ton of the enriched lead, while No. 9 will contain 2 tons of the impoverished lead. Each pot is then made up to 3 tons with lead containing the same amount of silver as that transferred to the respective pots, so that the same quantity of metal is operated upon in each case. The operation of melting, skimming and cooling, with constant stirring, is repeated in Nos. 9 and 7, when the lead is further enriched in the latter and impoverished in the former, so that when the portion travelling to the right reaches No. 15 pot it is nearly free from silver, while, when the portion which is transferred in the opposite direction reaches No. 1, it probably contains 250 ounces of silver per ton; but it may be concentrated to contain 700 ounces, beyond which it is not practicable to go.

"Parkes' process" depends on the superior affinity of silver for zinc over lead. The argentiferous lead is melted in an iron pot, raised to the melting point of zinc, skimmed, $1\frac{1}{2}$ to 2 per cent. of molten zinc added and the whole constantly stirred, while the temperature is gradually lowered until the zinc separates as a crust on the surface. These crusts, containing silver and some lead, are removed until an assay shows that the remaining lead is sufficiently low in silver. They are next heated to a little above the melting point of lead in a cylindrical retort open at the front, and inclining

CUPELLATION.

from back to front, when the lead liquates out, carrying much of the silver and a little zinc with it. The unmelted residue is then heated with lime and coal in a closed retort by which the zinc is distilled, leaving a mass of silver, lead, copper, etc., behind. The rich lead obtained by Parkes' or Pattinson's methods is treated by cupellation for the separation of the silver.

The desilverizing of lead is said to be facilitated by passing a voltaic current by means of copper wires through argentiferous lead containing a little zinc, until all the zinc has risen to the surface.

Cupellation.—This is a method for the separation of base metals as oxides in conjunction with oxide of lead, from silver and gold, which do not oxidize. Two methods are employed, known respectively as the German and the English.

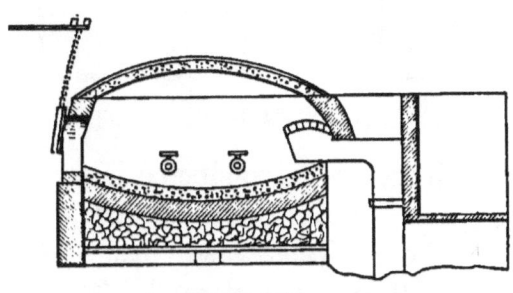

Fig. 52.

The "German hearth" (Fig. 52) is of the reverberatory type, but without a chimney, the volatile products escaping through an opening opposite the bridge. The bed, which is circular and concave and about 10 feet in diameter, is made of marl, which is chiefly composed of carbonate of lime, carbonate of magnesia and clay. The roof is formed of a movable iron cover lined with clay; the air is supplied by a pair of inclined twyers fixed in one side of the bed and the oxide of lead flows off through an opening opposite the twyers. When the lead is melted by means of a wood or coal fire, the blast is continued until the surface brightens, showing that the silver is comparatively pure and free from oxide. Water is then thrown on the fire and on the bath of silver, which soon solidifies. The silver thus obtained is further purified by melting in a small reverberatory furnace similar to the above, or in a plumbago crucible.

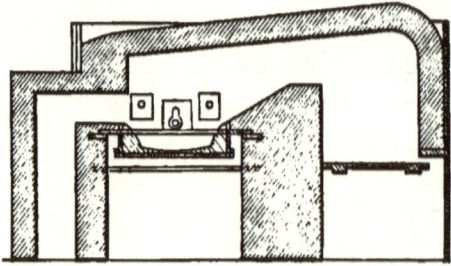

Fig. 53

The "English process" is conducted in a reverberatory type of furnace (Fig. 53), with a movable bed made of bone ash, called a "test." This is an oval frame of wrought iron (Fig. 54), about 5 feet long and 3 feet wide, well lined with moistened bone ash mixed with a little pearl ash, 1 inch thick, leaving a cavity which inclines from the sides to the centre, on which the metal is

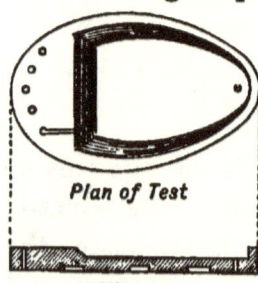

Plan of Test

Fig. 54.

afterwards melted and refined. The breast of this test is perforated with a number of holes for the removal of litharge as it forms. Some tests are kept in stock, as several weeks are required to gradually and thoroughly dry them after making. Bone ash differs from marl in that it absorbs a considerable portion of the litharge formed during cupellation.

Processes for the Extraction of Lead.

The various processes for extracting lead may be divided into three groups—1. By roasting and re-action. 2. By roasting and reduction with carbon. 3. By direct reduction with iron or bodies containing iron.

1. **Methods by roasting and re-action.**—These are carried out in reverberatory furnaces of which the Flintshire furnace (Fig. 55) may be taken as a type.

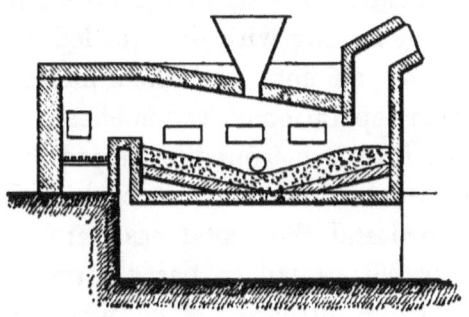

Fig. 55

The roof is low and inclines from fireplace to flue and contains in the middle a hopper for charging. Three working doors are arranged on each side, the front ones being used for rabbling the charge and the back ones for tapping the slag. The bed, which is lined with lead slag, is made to slope from all parts towards a depression called the well, in front of tap hole, which is placed under the middle door on the front side; in this well the metal collects before tapping.

The process is conducted in four stages—1. roasting; 2. reduction; 3. stiffening with lime; 4. tapping out the lead and removing the grey slag.

About 1 ton of galena is charged through the hopper and roasted for two hours on the raised part of the bed at a temperature below its melting point, with constant stirring to prevent clotting. Sulphurous acid is liberated and a mixture of sulphide, oxide and sulphate of lead remains. Then the fire is made up and the temperature raised until the charge melts; the re-action between the sulphide and oxides now takes place with the separation of metallic lead, which collects in the well. The mixture of slag and undecomposed ore is then stiffened by mixing with lime (called "setting up"); roasted again for an hour and then melted, by which means a further separation of lead is effected. The metal is now tapped into a pot outside heated by a separate fire, and the lead covered with fine coal. It is then well stirred, skimmed and the metal cast into moulds. The "grey slag" is withdrawn in pasty lumps through the back doors of the furnace. The whole operation of working off a charge lasts about five or six hours.

The "Brittany furnace" is smaller than the Flintshire, having three doors along one side only. The ore smelted is galena, with 12 to 18 per cent. of iron pyrites, and the method pursued is similar to that in North Wales. The lead obtained is purified by stirring with a pole of green wood as in copper refining.

In Spain a reverberatory furnace is employed for smelting rich galena by the method of roasting and re-action, but a kind of blank furnace is placed between the reduction chamber and the stack for the better control of the draught. Also the stiffening is made with charcoal instead of lime. In other respects it resembles the Flintshire method.

The "Cornish method," which is carried out in Cornwall and some continental works, is employed for the treatment of an impure galena containing blende, iron ore, fahlerz, quartz, fluor-spar, etc.

The reduction is effected partly by the principle of reaction and partly by iron and carbon, so that three products are obtained—lead, regulus which is again treated for lead, and slag. The ore, from $1\frac{1}{2}$ to 3 tons, is calcined in a separate furnace, then introduced into a furnace similar to the Flintshire furnace for reduction, and called the "flowing" furnace. A considerable quantity of lead separates by the re-action of the sulphide and oxides and is then tapped out. The residue is then mixed with anthracite and iron and the temperature raised, when a further reduction takes place yielding lead, regulus and slag. One operation in the flowing furnace takes seven to eight hours.

Method by roasting and reduction by carbon.—In the north of England, a small blast furnace is employed for lead smelting called the "ore hearth" (Fig. 56). It consists of a small rectangular chamber, 2 feet × 1 foot × 2 feet deep, lined with cast iron plates, and closed by an arched hood for taking away the fumes. In the back wall is fixed a twyer for introducing a blast of air. In front is an inclined work stone (*b*) with gutters for the lead to flow into the pot (*a*) placed in front of the work stone.

Fig. 56.

Peat, which is used as the fuel, is first charged into the hot hearth and the blast turned on. Then a little coal is added, and then the ore, which was formerly added in the raw state, but is now first roasted in a reverberatory furnace. In half an hour the lead begins to separate, when the contents are well stirred and a portion thrown on to the work stone *b*. The lead runs into the pot, the grey slag is separated and the residue returned to the hearth and mixed with lime; this is repeated at intervals, fresh fuel and ore being added as required. After some time, the cavity of the hearth becomes full of lead which flows along the gutters in the work stone into the lead pot. Two to three tons of lead are thus produced in about twelve to fourteen hours. This furnace is employed for poor ores where coal is scarce and peat plentiful.

Method by reduction with iron and carbon.— The method of smelting lead by roasting and reduction in blast furnaces at Freiberg is explained in the chapter on silver. The same principle is adopted in Sweden.

By the "Swedish process" the ore is smelted in two

EXTRACTION OF LEAD.

operations known as "raw" smelting and "lead" smelting. In the former, poor ores and various residues are smelted with iron pyrites, quartz, and lead slag in a blast furnace using charcoal fuel, to produce an argentiferous regulus and a clean slag.

In the "lead" smelting, rich ores and products, together with the roasted regulus from the "raw" smelting, some slag and a little roasted iron pyrites are treated, producing three products—1. argentiferous lead; 2. sulphide of iron; 3. slag, which is a silicate of iron and lime, etc. The reduction of the galena is effected by iron reduced from the roasted pyrites by means of carbon; the sulphur of the galena unites with the iron, and the remaining iron along with lime are fluxed by the quartz, thus:—

$$Fe_2O_3 + 3C = 3CO + Fe_2;$$
$$Fe + PbS = FeS + Pb;$$
$$FeO + CaO + SiO_2 = FeO, CaO, SiO_2.$$

At "Pontgibaud" a highly siliceous galena, rich in silver, is smelted. The ore is first calcined in a reverberatory furnace, 40 feet long × 15 feet wide, having six doors on each side. The ore is charged through a hopper at the flue end and gradually raked forward as the operation proceeds. The front part of the bed near the fire bridge is 6 inches lower than

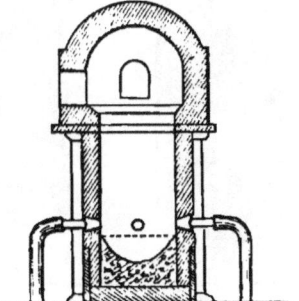

Fig. 57.

the other part; here the heat is sufficient to melt the charge, which is tapped at intervals of six hours. The roasting converts the galena into oxide and a little sulphate and the final melting forms some silicate of lead. The product is smelted in the Castilian blast furnace, shown in Fig. 57.

The furnace, constructed of blocks of lava bound with iron, is about 12 feet high × 3 feet wide, supported on iron pillars. The hearth contains three twyers.

The roasted and fused mass is mixed with 10 per cent. scrap iron, 16 per cent. limestone, and 3 per cent. fluorspar, together with some slags from a previous charge, and various residues. The products are,—impure argentiferous lead, 7 to 10 per cent. of regulus and slag. The reduction may be represented by the equation—

$$2PbO,SiO_2 + Fe_2 = 2FeO,SiO_2 + Pb_2.$$

In the "Hartz" and other parts of Germany, lead ore is smelted with basic silicate of iron in the Rachette furnace (Fig. 58). This furnace

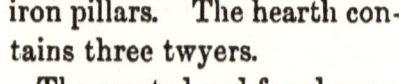
Fig. 58.

is oblong and rectangular in shape, being 3 feet wide at the twyers, 7 feet at the throat, and about 30 feet high, with a capacity of about

2000 cubic feet. The object of this shape is to keep the ascending gases more to the centre, and thus cause a more perfect action on the descending solid materials. The blast is introduced through six or eight twyers, three of four being placed on each of the longer sides of the hearth, thus causing a better distribution of the air. This furnace is also used in Russia for iron smelting, and for reducing copper slags in the Lake Superior district. The re-actions may be represented thus—

$$2FeO,SiO_2 + PbS = PbO + FeO,SiO_2 + FeS$$
$$2PbO + C = Pb_2 + CO_2$$

from which it will be seen that the products are lead, regulus and slag.

Electrolytic method.—Lead may be deposited electrolytically in a coherent state by operating on its alkaline solutions, preferably those containing phosphoric and tartaric acids. Galena must be roasted to oxide of lead, which is readily soluble in caustic potash and soda.

Lead fume.—The gases and vapours arising from lead smelting furnaces contain considerable quantities of lead, which are recovered by condensing the fume in suitable flues or condensers. The most common method is to construct slightly inclined brick flues, which wind about in a serpentine manner, generally up the sides of a hill, the whole course extending in some cases for several miles before the vapours escape into the atmosphere. Doors are fixed at intervals for purposes of cleaning out the condensed matter.

Another class of condenser consists of a divided chamber, containing water at the bottom, through which the vapours are drawn several times by means of a pump or other exhauster. Fig 59 shows Stagg's arrangement.

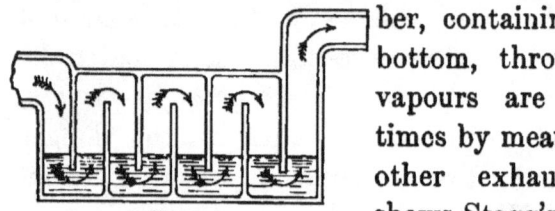
Fig. 59.

Lead fume consists of lead sulphide, oxide, carbonate and sulphate, and sometimes silver, together with carbonaceous matter, lime, alumina, oxide of iron, silica, etc.

This condensed fume is treated with slags and other products in the slag hearth or furnace (which is generally a small cupola blast furnace), producing an impure lead known as "slag lead." The reduction of the lead compounds and the scorification of the impurities is effected by means of coal or coke and iron slags. The lead obtained is hard owing to the presence of sulphur, antimony, copper, iron, etc., and requires to be purified: this is effected in a reverberatory furnace, known as the "lead softening furnace."

The softening furnace is a low reverberatory one; the bed contains a shallow cast iron pan, in which 10 tons of hard lead are melted and exposed to an oxidizing flame, whereby the antimony, arsenic, copper, iron, etc., with some lead are oxidized and form a dross on the surface. This is periodically raked off and the refining continued until a sample of the lead on trial shows the requisite malleability. The operation requires from eighteen to twenty-four hours. In some works a blast of air is used to assist the oxidation.

Reduction of oxidized compounds.—Litharge, produced from cupellation, and skimmings from Pattinson's pots, are reduced with small coal in a reverberatory furnace similar to the Flintshire furnace, or in a cupola, such as that used for reducing lead slags.

Manufacture of Red Lead (Pb_3O_4).

This oxide is prepared on the large scale by heating about 1 ton of lead in a reverberatory furnace so as to form massicot (PbO), known as the drossing stage, and afterwards colouring. The dross as it forms is scraped from the surface, thus exposing the molten lead to further oxidation, the whole being converted in about twenty-four hours. The massicot is allowed to cool, raked out, moistened with water and ground under edge stones to a fine powder. It is then washed with water, and carried forward in suspension by a stream of water, from which it is deposited in settling tanks.

The next stage, called colouring, consists of placing dried massicot upon the bed of a reverberatory furnace, and carefully heating at a temperature of about 300° C. for two days, with frequent stirring, until the whole appears of a dark red colour, which changes to bright red on cooling; the furnace is then closed and allowed to slowly cool. The product is red lead, which, after grinding and sifting, is ready for use. Red lead is frequently adulterated with oxide of iron and brick dust, which may be detected by fusing a small portion, when the red lead is converted into yellow litharge, while the impurities retain their red colour. Red lead is reduced when heated with carbon thus—

$$Pb_3O_4 + C_2 = 2CO_2 + 3Pb.$$

White lead is a basic carbonate of lead, or a mixture of carbonate and hydrate in variable proportions, of which the following formula may be taken as a general guide— $2PbCO_3, PbH_2O_2$. It is prepared by exposing metallic lead to the action of moist air and carbonic acid. It is often adulterated with heavy spar, gypsum and zinc oxide.

CHAPTER IX.

TIN (Sn).

TIN is a white metal with great lustre, very malleable, but very low in tenacity. It melts at 230° C. and may be strongly heated without volatilizing. A bar of tin when bent produces a crackling sound known as the "cry," which is a rough test of its relative purity. Its specific gravity is 7·25. It is little affected by air, but rapidly oxidizes when heated, forming stannic oxide (SnO_2). When melted and poured at a temperature near its solidifying point the surface remains bright, if pure, but the presence of a little lead, iron, etc., imparts a dull or frosted appearance. The surface of tin or tin plate can be readily crystallized by treating it with a mixture of 10 parts dilute sulphuric and 1 part of dilute nitric acid, the appearance being known as "moirée métallique." Commercial tin often contains small portions of lead, iron, copper, arsenic, antimony, bismuth, tungsten; sometimes manganese and zinc.

Alloys of tin.—Tin plate is iron with a coating of tin on its surface, which will be brighter the purer the tin. Terne plate is an inferior quality, containing a little lead. Bronze contains 80 to 95 per cent. copper and 20 to 5 per cent. tin; bronze coinage contains 95 copper,

4 tin, and 1 zinc; gun metal, 80 to 90 copper and 20 to 10 tin; bell metal, 70 to 80 copper and 30 to 20 tin; speculum metal, 2 copper, 1 tin, and a little arsenic; for pewter and soft solder, see Lead; for Britannia metal and type metal, see Antimony; for fusible alloy, see Bismuth.

Tinning iron plates.—The iron plates are first thoroughly cleansed in warm dilute sulphuric acid, then washed and scrubbed with sand to remove all traces of rust, which would prevent the tin from adhering; the plates are then immersed in a bath of melted tallow, which dries them thoroughly.

The tinning arrangement comprises five pots, each heated by a separate fire, and termed respectively—the tin pot, washing pot (divided into two compartments), grease pot, cold pot and the list pot. The plates from the tallow or tinman's pot are placed in the tin pot, the surface of the molten tin being covered with tallow to prevent oxidation; after being heated for an hour and a half in the melted metal they are removed, drained and plunged into the first division of the washing pot. This also contains molten tin; then they are removed and brushed to remove excess of tin and afterwards quickly dipped into the tin in the second division to remove the brush marks. Then they are transferred to the grease pot containing melted tallow, which removes any excess of tin. After about ten minutes the plates are inserted in melted tallow contained in the cold pot. Lastly, in order to remove the tin which has drained down and formed a bead at the lower edge of the plate, each plate is placed in the list pot, which con-

EXTRACTION OF TIN.

tains melted tin about $\frac{1}{4}$ inch deep; by striking the plate sharply with a stick the superfluous metal is detached. The plates are then rubbed with bran and afterwards with sheepskin, when they are ready for use.

An inferior variety of tin plate, in which the iron is coated with an alloy of tin and lead, is termed "terne plate."

Ores of tin.—The only important ore of tin is tinstone (SnO_2), which is usually of a black or brown colour, and sometimes grey, with a specific gravity of 6 to 7. It is largely found in Cornwall, more or less mixed with wolfram, fluor-spar, granite, galena, blende, pyrites, etc. It is sometimes found in masses with concentric layers, and is then called "wood tin," sometimes in grains with sand called "stream tin," but generally in mineral veins called "mine tin."

METHODS OF EXTRACTION.

Tin stone is smelted both in blast and reverberatory furnaces.

The reverberatory method is conducted in two stages. Ores containing arsenic, sulphur, etc., are first roasted to expel these impurities and convert the metals into oxides and sulphates, the latter being afterwards dissolved out in water.

The reduction is effected in an ordinary reverberatory furnace with a low slanting roof; the bed, about 18 feet by 9 feet, is hollowed out, all parts inclining towards the tap hole. The bottom is formed of iron, covered with slate, and lined internally with 7 to 8 inches of fire-clay. 24 to 30 cwts. of ore are mixed with $\frac{1}{5}$th their weight of anthracite, a little lime and fluor-spar being added as a flux. In five hours the mass is well rabbled, more anthracite is added and in about another hour the metal is ready for tapping. The slag is a silicate of iron containing other oxides.

The metal reduced in the manner just described is very impure and requires to be refined. The pigs of tin are placed on the bed of a reverberatory furnace and gradually raised to the melting point of tin, when the tin liquates out, leaving a mass known as "hard head," which contains about 50 per cent. iron, 20 per cent. tin, 20 per cent. arsenic and a little sulphur, copper, etc. The liquated metal is melted in an iron pot and stirred with a pole of green wood; this causes a boiling, due to the escape of gases, and brings the impurities—iron, arsenic, etc., to the surface. These are skimmed off from time to time. Instead of poling, the metal is sometimes raised in ladles and allowed to fall from a height into the pot. This is called "tossing."

For "common" tin, the metal is ladled from the refining pot into moulds. For "refined" tin, purer ores are employed for its extraction and the poling operation continued longer. The metal is then allowed to settle. The upper portion being the purest is preserved for "grain" tin, the middle portion for common tin, and the bottom layers require to undergo another liquation and poling. The moulds used for casting are made of granite and the ingots are known as "block" tin.

The peculiar structure of grain tin is produced by dropping the metal from a height when on the point of melting.

Blast furnace method.—In Germany, tin ore is smelted in a small charcoal blast furnace (Fig. 60). It is built of granite and covered with a hood at the top. The metal and slag flow into the receptacle (A) and the liquid slag on the surface flows along a gutter into a vessel (B) containing water, in order to make it brittle, so that it may be easily broken up for re-smelting to recover the tin which it contains.

Fig. 60.

Deposition of tin.—Tin is deposited on articles of brass, copper, or bronze, by immersing them in a boiling solution of peroxide of tin in caustic potash, in contact with cuttings of metallic tin. The same solution may be used for tinning iron with the aid of a battery, using a large tin anode. Large crystals of tin may be formed by sending a strong current of electricity through a solution of stannous chloride, using a small anode.

CHAPTER X.

Nickel (Ni).

Nickel is a brilliant white, malleable, ductile, weldable and very tenacious metal, with a melting point only a little below that of iron, but the presence of carbon and other impurities considerably lowers this point. Its specific gravity is 9 and it is magnetic like iron, although in a much less degree. It does not readily oxidize in air, but when heated it forms nickel oxide (NiO). It readily unites with sulphur, forming NiS, which is brass-yellow in colour, and with arsenic to form Ni_2As. When nickel is fused with carbon, the two elements combine to form a compound more fusible and brittle than pure nickel, in proportion to the amount of carbon taken up; in this respect it resembles iron.

Nickel speise is a yellowish white, brittle, metallic looking substance, consisting of nickel and arsenic with sulphur, iron, cobalt, etc., and occasionally contains antimony and copper. When heated in air or oxygen, the elements are oxidized in the order of their oxidizability —sulphur and iron first, then cobalt, then nickel.

Alloys of nickel.—The chief alloy of nickel is German silver, or nickel silver, which consists of copper,

EXTRACTION OF NICKEL.

zinc and nickel in various proportions and sometimes a little iron, thus :—

Zinc.	Iron.	Nickel.	Copper.
25·4	2·6	31·6	40·4
13·6	—	19·3	67
22·15	--	15·45	62·4
26·55	—	10·85	62·60

Ores of nickel.—Kupfer nickel or copper nickel ($NiAs_2$) is a copper-red coloured mineral with a metallic lustre. Nickel pyrites (NiS) is brassy-yellow in colour. Nickel-glance is a variable compound of nickel, arsenic and sulphur. Garnierite is a hydrated silicate of nickel, iron, and magnesium.

METHODS OF EXTRACTION.

The extraction of nickel from its ores consists of three stages—1. Concentration by roasting to form regulus or speise. 2. Conversion of regulus or speise into oxide by dry or wet methods. 3. Reduction of nickel oxide by carbon.

The extraction is based on the following principles, cobalt being generally present :—

1. When sulphides or arsenides of nickel and cobalt are fused with an acid silicate of iron, the cobalt passes into the slag, but the nickel does not leave the regulus or speise.

2. When oxides of nickel and cobalt are smelted with sulphur or arsenic, the nickel unites with the sulphur or arsenic, but the cobalt only partially.

3. When silicates of nickel and cobalt are fused with sulphur or arsenic, the nickel forms a regulus or speise, but the cobalt does not leave the silicate.

Dry method.—The dry method of extracting nickel is very similar to the process for extracting copper, consisting of alternate roastings and meltings, in which advantage is taken of the superior affinity of nickel for sulphur and arsenic (as shown in the preceding principles), by which it is preserved from oxidation, while the more oxidizable metals are being removed.

Wet method.—The wet method is the one most generally adopted, being necessary when much cobalt is present in the ore. The scheme Fig. 61 represents one mode of treating nickel ore—

1. The ore is ground and calcined to remove arsenic and sulphur as far as possible; 2. dissolved in hot hydrochloric acid; 3. bleaching powder is added to peroxidize the iron, which is precipitated as basic arseniate of iron; 4. the filtrate is treated with sulphuretted hydrogen, which precipitates copper as sulphide; 5. the filtrate containing nickel and cobalt is boiled to expel sulphuretted hydrogen and neutralized with lime; 6. oxide of cobalt is precipitated with bleaching powder and the solution filtered; 7. oxide of nickel is precipitated by boiling with milk of lime; 8. nickel oxide is reduced by carbon.

Oxide of nickel is sometimes reduced by making it into a paste with carbon, cutting it into cakes or cubes, placing these in crucibles or tubes and raising them to a white heat while surrounded with charcoal.

Nickel plating.—Nickel is largely used in electro

EXTRACTION OF NICKEL. 155

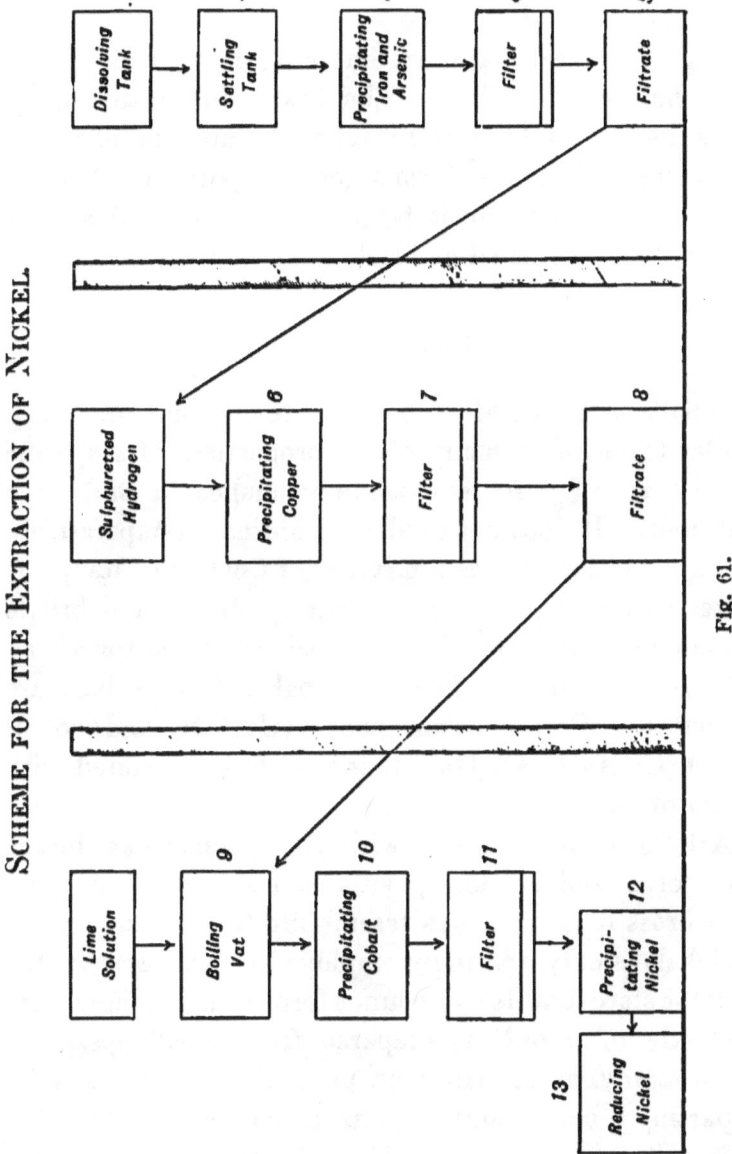

Fig. 61.

plating, its great hardness enabling it to resist wear much better than silver. The best solution to employ is the double sulphate of nickel and ammonia, of which 12 ounces to the gallon form a good proportion. Articles to receive a deposit must be perfectly clean and should be previously polished as brightly as possible.

Cobalt (Co).

Cobalt is a metal of a steel-grey colour and very similar to nickel in many of its properties. Its specific gravity is 8·9; it is extremely malleable and very tenacious. It does not oxidize at ordinary temperatures, but at a red heat forms a mixture of CoO and Co_3O_4. It unites with arsenic to form iron-grey, fusible and brittle compounds, when arsenic and cobalt are fused together.

Ores of cobalt.—The principal ores of cobalt are —smaltine ($CoAs$), cobalt-glance (Co_2AsS) and cobalt bloom ($Co_3AsO_4, 4H_2O$). It is also often associated with nickel ores.

Alloys.—Cobalt forms with tin a somewhat ductile alloy of a violet colour; with iron a hard alloy; and with brass it forms alloys resembling German silver.

Cobalt is only sparingly employed in the arts in the metallic state, but its compounds form valuable pigments.

Oxide of cobalt is prepared from cobalt speise by the wet way, as described on page 154, and is used for imparting a blue colour to glass, enamel and pottery.

Smalt.—This is a double silicate of cobalt and potassium, also called cobalt glass. Its preparation is conducted in two stages—1. the ore is calcined to expel arsenic sulphur, etc.; 2. fused with sand and carbonate of potash, when it forms a deep blue glass.

Zaffre is an ore of cobalt containing sufficient silica to form smalt when fused with carbonate of potash.

Printers' blue is used for painting the blue colour on pottery and china, etc. It is prepared by fritting a mixture of silica, oxide of cobalt, nitre and a little barium sulphate.

CHAPTER XI.

Aluminium (Al).

This is a white metal which takes a fine polish. It has no taste or odour. It is soft, very malleable and ductile, with an elasticity and tenacity about equal to that of silver. Its specific gravity is 2·5, which is increased by hammering; it melts at a temperature a little above that of zinc, and is not volatile when strongly heated out of contact with air. Its conductivity for heat and electricity is very high, as is also its specific heat. It does not oxidize in air or combine with sulphur; it is insoluble in cold nitric acid; sulphuric acid has no action on it, but hydrochloric acid and alkalies dissolve it readily.

Ores of aluminium.— Aluminium occurs in a variety of forms as oxide, but more generally in combination with other metals, such as zinc, iron, magnesium, etc., forming aluminates. It exists as silicate in all clays. Also as fluoride in cryolite ($6NaF, Al_2F_6$).

Methods of Extraction.

The general method depends on the action of sodium at a red heat upon the chloride, fluoride, or double fluoride of sodium and aluminium.

EXTRACTION OF ALUMINIUM. 159

10 parts of the double chloride of aluminium and sodium, 5 parts of fluor-spar and 2 parts of sodium are heated in a closed reverberatory furnace, when a vivid action ensues, evolving great heat. When the reduction is complete, the main portion of the slag is run off first, and then the aluminium which collects at the bottom of the receptacle under a layer of liquid slag. The upper and lighter portion of the slag consists chiefly of common salt, the heavier and more infusible portion is chiefly aluminium fluoride containing shots of metal.

Cryolite which occurs abundantly in West Greenland is the most valuable ore of aluminium and is extensively used as a source of the metal. The powdered mineral is mixed with half its weight of common salt and the mixture arranged in alternate layers with sodium, using 2 parts of sodium to 5 parts of cryolite.

Aluminium may be freed from copper and iron by fusion with nitre, and from zinc by volatilization of the latter. Slag may be removed by melting the metal in a blacklead crucible and stirring with an oxidized iron rod.

The cheap production of aluminium depends to a large extent on the price of sodium. This metal has lately been obtained by a new process introduced by Castner, which it is stated reduces the cost of production from four shillings to one shilling per pound. The charge consists of caustic soda, mixed with a mixture of iron and carbon, obtained by coking pitch with iron filings. The iron makes the particles of carbon heavier and keeps them from floating to the surface, so that the soda is more perfectly reduced and a lower temperature can be employed with a great saving of fuel. The operation is conducted in large crucibles.

Aluminium may be obtained by electrolysis. The double chloride or fluoride of aluminium and sodium is fused at a high temperature in a crucible divided into two parts by a porous diaphragm. A strong current is sent through the liquid, using carbon or platinum electrodes, when the metal is deposited at the negative pole.

The Cowles Electric Smelting Co. have succeeded in reducing aluminium compounds by carbon, or by dissociation assisted by carbon, in an electric furnace. The interior of the furnace is lined with carbon, and through this runs a core of carbon blocks which are traversed by an electric current from a powerful dynamo. The carbon is raised to an intense heat, and if oxide of aluminium in the form of corundum, is mingled with the carbon, aluminium is separated; but some of the metal is volatilized.

Alloys.—Aluminium bronze consists of 90 parts copper and 10 parts aluminium. It has a gold colour, takes a high polish, is very hard and malleable and has a tenacity of 48 tons per square inch. The addition of 2 to 3 per cent. of brass increases its tensile strength and renders it less liable to oxidation. The addition of nickel to aluminium bronze also increases its strength. When more than 10 per cent. of aluminium is present with copper, the malleability is diminished and 15 to 20 per cent. forms a brittle alloy.

Silver, iron, and zinc alloyed with aluminium form hard and brittle alloys.

Alloys of aluminium may be prepared by heating

alumina with carbon and some metal such as copper. In this way aluminium copper is made in the Cowles electric furnace, which alloy is afterwards employed in making aluminium bronze.

Aluminium unites with mercury when mixed with sodium amalgam, or when dipped in a potash solution containing mercury.

CHAPTER XII.

MERCURY OR QUICKSILVER (Hg).

MERCURY is silver white with a brilliant lustre. It is the only metal liquid at the ordinary temperature; at 360° C. it boils; its specific gravity is 13·6; it has a high and fairly regular coefficient of expansion for heat, which renders it suitable for thermometers and similar instruments. It does not oxidize in air except near its boiling point, which forms a ready means of detecting the presence of base metals that exist in it as impurities or adulterations.

Vermilion.—Mercury forms an important compound with sulphur, known as vermilion (HgS), which is prepared by gradually adding 200 lbs. of mercury to 32 lbs. of molten sulphur in an iron pot, with constant stirring; the mass is then poured on to an iron slab and broken up, then placed in earthen crucibles and sublimed. Vermilion has a beautiful red colour, and sublimes unchanged when heated out of contact with air. Heated with access of air it is decomposed, mercury and sulphurous acid being formed. It is also decomposed when heated with carbon, iron and most metals, lime, alkalies and hydrogen.

Amalgams.—Mercury unites with most metals forming "amalgams," some of which are liquid, others semi-liquid and some solid. The solid amalgams are regarded as chemical compounds, while the liquid amalgams may be solutions of compounds in excess of mercury. But the affinity is feeble, as the mercury is partially expelled by pressure, and completely so by heat in most cases. Amalgams are formed—1. by rubbing the metal in a finely divided state with mercury, an increase of temperature facilitating the amalgamation; 2. by dipping the metal into the solution of a mercury salt; 3. by voltaic action, as when a metal is placed in contact with mercury and an acid; 4. by mixing a metal such as gold with an amalgam of a highly positive metal such as sodium.

An amalgam of mercury with tin is used in silvering mirrors; with gold and silver in water gilding, and with tin and gold or platinum in dentistry, etc. Sodium amalgam is prepared by rubbing the two metals together in a covered mortar. An amalgam of 30 parts mercury to 1 part sodium is solid. It is used for amalgamating iron and platinum and in the extraction of gold and silver from their ores.

Ores of mercury.—Mercury sometimes occurs native, mixed with other of its ores or with gangue, or as an amalgam with silver; occasionally as chloride, bromide and iodide of mercury. The chief source of the metal is the sulphide (HgS), known as "cinnabar." It is a brownish red mineral, moderately soft and gives a scarlet streak when drawn across unglazed porcelain.

Modes of Extraction.

Mercury being a volatile metal is obtained by a process of distillation like zinc, but a far more extensive condensing arrangement is required. The various methods may be classed under two heads—1. air reduction; 2. reduction by iron or lime.

1. Air reduction.—The mode of decomposition may be represented by the equation $HgS + O_2 = SO_2 + Hg$. At Almaden, in Spain, the reduction chamber is a "dome" furnace (see Fig. 6), with an opening near the top communicating with a system of earthenware condensers called "aludels" (Fig. 62).

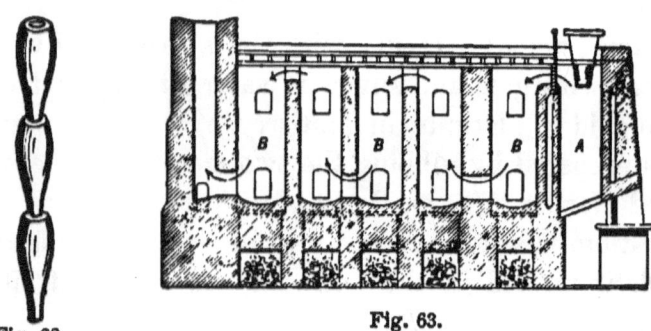

Fig. 62.   Fig. 63.

The Hähner furnace (Fig. 63) consists of a vertical cylinder (A), 18 feet high and 3 feet 8 inches in diameter, into which the ore is fed from a hopper; six condensing chambers (BB), the roofs of which are covered with an iron plate, which forms the floor of a tank, through which water is kept circulating. The vapours are compelled to pass in the direction of the arrows, the communication between the chambers being alternately at top and bottom. The current is maintained by a chimney draught. The chimney is built in three tiers, in each of which are ledges kept cool by water flowing over them so as to complete the condensation. The charge consists of 7 cwts. ore with 5 per cent. of charcoal. The furnace is worked continuously.

The Alberti furnace (Fig. 64) is employed for "smalls"

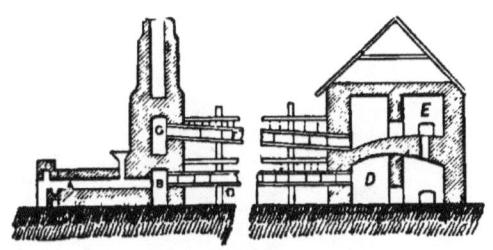

Fig. 64

and poor ores, containing about 1 per cent. of mercury. It consists of a reverberatory furnace (A), capable of holding 3 tons, with a sloping bed arranged in three steps, the ore being gradually pushed forward from the flue end to the fire bridge. The vapours pass into a condenser (B), and thence along the iron tubes (C), then into a condenser (D), up into (E), then back through the tubes (F) into a chamber (G) at the base of the chimney, which is obstructed by a number of partitions, around which the vapours pass so as to condense any remaining mercury. The iron pipes (C and F) are kept cool by running water.

Closed retorts have also been employed for mercury ores using lime, iron, or oxide of iron as the reducing agent, when the following re-actions occur —

$$4HgS + 4CaO = 3CaS + CaSO_4 + 4Hg$$
$$HgS + Fe = FeS + Hg$$
$$5HgS + Fe_3O_4 = 3FeS + 5Hg + 2SO_2.$$

Mercury is purified from zinc, bismuth, antimony, and tin, which form its chief impurities, by placing it in a retort, covering with iron filings and distilling the mercury, which is condensed by the leading pipe dipping into water.

CHAPTER XIII.

Antimony (Sb).

Ordinary commercial antimony is very impure, containing iron, lead, arsenic, and sulphur and is called "regulus of antimony." Antimony is a brilliant bluish white metal, highly crystalline, with fern-like markings on the surface and very brittle, so that it may be easily powdered; its specific gravity is 6·8; it melts at 450° C., and volatilizes at a higher temperature. It does not oxidize at ordinary temperatures, but when heated in air, antimonious oxide (Sb_2O_3) is formed, and at a red heat it burns with a bluish white flame producing dense white fumes of Sb_2O_3.

Antimony and sulphur readily unite when heated together forming Sb_2S_3; the same compound is also formed by heating the oxide with sulphur, thus—

$$2Sb_2O_3 + 9S = 3SO_2 + 2Sb_2S_3.$$

When this sulphide is roasted in air or fused with nitre, the oxide is re-formed. It is also decomposed by fusion with carbon, iron, tin, copper, potassium cyanide and a mixture of carbonate of soda and charcoal. Sb_2S_3 is a bluish grey crystalline compound, with a metallic lustre.

Alloys of antimony.—Antimony unites with other metals to form valuable alloys in consequence of its hardening properties, but it impairs the malleability and ductility of the malleable metals. Type metal sometimes contains 17 to 20 per cent. of antimony and 83 to 80 of lead, and sometimes 50 lead, 25 tin, and 25 antimony. For stereotype plates, 70 lead, 15 antimony and 15 bismuth are used; and for music plates, the alloy contains 12 tin, 7 lead, and 1 antimony. Britannia metal has a variable composition, but the following may be taken as a guide, 90 tin, 7 antimony, and 3 copper. A violet coloured alloy known as "regulus of Venus," contains 50 copper and 50 antimony. The effect of even small quantities of antimony on the malleable metals, such as copper, gold, iron, etc., is most injurious, making them hard and brittle.

Ores of antimony.—Antimony occurs native and in combination with other ores, but the chief ore is "Stibnite" (Sb_2S_3).

METHODS OF EXTRACTION.

The reduction of stibnite is performed in two stages— 1. liquation, by which the sulphide is separated from its gangue; 2. reduction of the sulphide by means of iron, carbon, etc.

The liquation is effected in the Hartz by placing the ore in conical pots covered at the top, perforated at the bottom, and standing on receivers sunk in the ground; the space between the pots is occupied by the fire. The fused sulphide runs through the perforations, and collects in the receivers placed below. At Malbose in France, the ore is placed in large cylinders (A) (Fig. 65), perforated at bottom, each holding 500 lbs., four being placed in one furnace, and standing on perforated plates (B) beneath which are chambers containing earthen receivers for the liquated material. At Linz, the liquation is performed in a reverberatory furnace, the bed inclining towards the tap hole through which the liquated sulphide flows into a receiver placed outside.

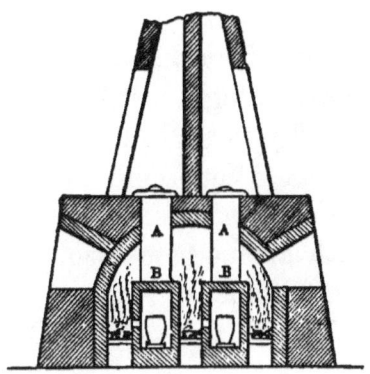

Fig. 65.

The reduction of the sulphide in England is performed in large crucibles heated in circular wind furnaces. Each crucible is charged with 40 lbs. of sulphide and 20 lbs. of scrap iron, the product being impure antimony and sulphide of iron, thus—

$$3Fe + Sb_2S_3 = 3FeS + Sb_2$$

The crude metal is then melted in crucibles with sulphate of soda and a little slag from the next process, the charge consisting of 80 lbs. metal, 2 lbs. salt cake and a little slag.

The metal thus obtained is melted with pearl ash and some slag, and cast into ingots for commerce. With rich ores the preliminary liquation is unnecessary.

Arsenic (As).

This metal has a steel-grey colour and metallic lustre; is exceedingly brittle and crystalline; its specific gravity is 4·7; when heated to 180° C., it passes into vapour without liquefying, but may be liquefied when heated under pressure. Heated in air it oxidizes and burns with a blue flame evolving a garlic odour and forming white fumes of As_2O_3.

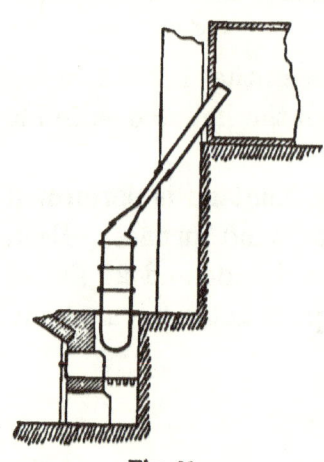

Fig. 66.

The white oxide (As_2O_3) called arsenious acid exists in the amorphous and crystalline forms. It is obtained from the calcination of many ores containing arsenic, forming transparent masses called arsenical glass. It is purified by re-subliming in deep iron vessels (Fig. 66), and collecting the vapours in a divided condensing chamber. When heated with carbon, carbonic oxide, or hydrogen below redness, it is reduced to metallic arsenic, thus—

$$2As_2O_3 + 3C = 3CO_2 + 4As.$$

Arsenic forms two compounds with sulphur—realgar (As_2S_2) and orpiment (As_2S_3).

Yellow orpiment or king's yellow (As_2S_3) is prepared by subliming a mixture of 7 parts (As_2O_3) with 1 part sulphur.

Realgar is an amorphous, brownish red substance, obtained by fusing 1 part of arsenic with 2 parts of sulphur or 2 parts of As_2O_3 with 7 parts of sulphur.

Ores of arsenic.—Arsenic occurs in nature as realgar, orpiment, mispickel (FeAs + FeS), nickel pyrites (NiAs), and kupfer-nickel or copper nickel ($NiAs_2$).

Method of extraction.—The metal is obtained by heating nickel pyrites, mispickel, etc., in closed retorts, when the arsenic is expelled and sublimes into condensing chambers.

Alloys of arsenic.—Arsenic enters into the composition of some alloys such as shot metal, its general effect being to harden and render the alloys brittle and more fusible. Its compounds are used in medicine and in glass making.

BISMUTH (Bi).

Bismuth has a greyish white colour with a tinge of red and bright lustre; it is brittle and crystalline; it fuses at 270° C., and volatilizes at a high temperature, burning with a blue flame, forming "flowers of bismuth" (Bi_2O_3). It expands on cooling, so that the metal is denser in the liquid than in the solid state; its specific gravity is 9·9, which is reduced by pressure; it oxidizes in moist air, and may be used in place of lead in cupellation. It unites with sulphur, forming a dark grey metallic looking sulphide (Bi_2S_3).

Ores of bismuth.—It occurs in the metallic state as bismuth-glance (Bi_2S_3), as ochre (Bi_2O_3), and often accompanies ores of silver, lead, tin, copper and cobalt.

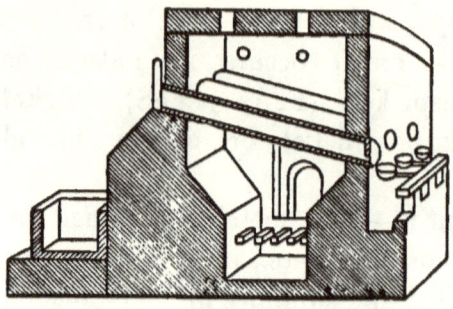

Fig. 67.

Method of extraction.—Bismuth is separated from several of its ores by liquation. The furnace employed is shown in Fig. 67. It contains a series of iron tubes inclining towards the front, in which the ore is placed. The fireplace is below, wood fuel being employed. The liquated metal is received in iron pots kept hot by a separate fire. It is purified by re-melting with nitre.

Alloys of bismuth.—Bismuth is used chiefly in alloys to render them easily fusible; they also expand on solidifying, especially the alloys of bismuth with antimony. Fusible alloy consists of 1 part lead, 1 of tin and 2 of bismuth; the addition of a little cadmium makes it still more fusible; this is known as "Wood's alloy," and melts at 82° C.

APPENDIX.

QUESTIONS ON METALLURGY.

NOTE.—*Questions marked * are of an advanced character, and therefore beyond the scope of the present work.*

PHYSICAL PROPERTIES OF METALS.

1. What is meant by the word 'Metallurgy'?

2. What is a metal? Name those of importance in the Arts, and state which are unfit for use in the metallic state.

3. State the most characteristic physical properties of metals, and explain what you understand by the term 'physical properties.'

4. Define the terms 'lustre' and 'colour,' as applied to metals. Give instances in which the colour and lustre respectively are modified by small quantities of impurity.

5. Give instances of the influence of small quantities of impurity on the physical properties of metals.

6. Mention the names of the metals in common use in the order of their fusibility, and give approximately the temperature at which each begins to fuse.

7. What metal passes, when heated under ordinary atmospheric pressure, directly into the gaseous state?

8. Express approximately in degrees Centigrade the temperatures represented by the terms—'incipient red,' 'dull red,' 'bright red,' 'white' and 'dazzling white' heats.

9. Some metals are said to be 'fixed,' others are said to be 'volatile' by heat. Explain these expressions.

10. Do all metals liquefy directly on heating ? If not, name any exceptions.

11. Name some metals which are lighter than water. How is the relative weight expressed and how is it determined ?

12. What is the effect of rolling and wire-drawing on metals ? If copper be rolled out cold, will its specific gravity be increased or diminished ? Give reasons for your answer.

13. Describe the different kinds of fracture in metals, and give an example of each kind.

14. Give instances in which metals may be separated from each other by crystallization. What conditions are most favourable to the formation of good crystals ?

15. What is meant by the term 'tensile strength' and how is such strength ascertained ?

16. What conditions are most favourable for producing high tenacity in a metal and what conditions are unfavourable ? Give a list of metals in the order of their tenacity.

17. What is meant by 'toughness' and how does it differ from tenacity ?

18. What is understood by a soft metal ? How is a soft metal generally hardened ?

19. Explain the term 'malleability' and give a list of metals in the order of their malleability.

20. Mention the various conditions which modify the malleability of a metal.

21. What is meant by the word 'annealing'? State why metals are sometimes annealed.

22. If you wished to render (1) copper and (2) lead hard, how would you proceed?

23. Explain the term 'ductility.' What physical properties conduce most to render a metal ductile?

24. How is the ductility of zinc affected by varying temperatures below a red heat?

25. What is meant by the 'elasticity' of a metal? How would you determine the degree of elasticity of a piece of steel?

26. Give a clear definition of the term 'capacity for heat,' and arrange the metals iron, nickel, lead, tin, bismuth, silver, copper and zinc in the order of their capacity for heat.

27. Do all metals expand on heating and contract on cooling? Do you know any exceptions?

28. A wooden rail feels warmer than an iron one, although both may be at the same temperature. Explain this.

29. What is meant by the term 'conductivity'? Give a list of metals in the order of their conductivity.

30. If a spiral of copper wire is placed in a candle flame the flame is extinguished. Explain this.

31. Explain the term 'welding.' What metals can be welded cold? State the conditions necessary for success in welding.

32. What is the result in each case of submitting finely-divided gold, zinc and lead to strong compression?

33. What is meant by the term 'autogenous soldering'?

34. What is an alloy? Name some metals which alloy well together and others which do not.

35. What reasons are there for supposing some alloys are true chemical compounds and others mere mixtures?

36. State the composition of useful alloys in which tin, antimony and zinc occur.

37. State approximately the composition of—brass, bronze, German silver, pewter and speculum metal.

38. Give approximately the percentage composition of the following alloys:—Muntz-metal, electrum and type metal.

39. State the composition of the following alloys:—Bell-metal, yellow metal, soft solder and fusible alloy.

40. What is the composition of:—Delta-metal, Aich-metal, aluminium bronze, sterro-metal and nickel-silver?

41. What do you understand by a homogeneous alloy? How is the homogeneity of some alloys affected by the rate of cooling?

42. What is the composition of the British gold, silver and bronze coinage?

43. Mention some very hard alloys and state for what purposes each is used.

Metallurgical Terms.

44. In what condition do metals generally occur in nature? Explain the term 'native' as applied to an ore.

45. Explain the terms 'matrix,' 'vein-stuff' and 'gangue.'

46. How does an ore differ from an alloy?

47. What is meant by the term 'dressing'? Mention cases in which this operation is unnecessary.

48. Define the terms 'smelting,' 'reduction' and 'oxidation.'

APPENDIX.

49. What is meant by the terms 'calcination' and 'roasting'? Illustrate by an example in each case.

50. What takes place when a mixture of red lead and charcoal is heated to redness in a covered crucible and what word should be used to express that action?

51. What kind of furnace is generally employed for roasting ores?

52. What is a flux? By what principles would you be guided in the choice of a flux?

53. What is a slag? What are the general features of a good slag?

54. Of what substances are slags generally composed? On what does the fusibility of a slag depend?

55. A slag is sometimes 'basic,' sometimes 'acid' and sometimes 'neutral.' Explain these terms and give examples of each kind of slag.

56. What is a scoria and how does it differ from a slag?

57. How does fluor-spar act as a flux, and for what substances is it chiefly employed?

58. What occurs if you heat a mixture of sand and lime in such proportions that the former contains twice as much oxygen as the latter?

59. Can you name any metals which are reduced from their ores by heat alone and any by the combined action of heat and air?

60. What are principal reducing agents used in metallurgical processes?

61. Explain fully the action of carbon in the reduction of metallic oxides.

62. Give a definition of the terms 'distillation' and 'sublimation.' Give illustrations of each in the extraction of metals.

M

63. In what metallurgical processes is mercury used? Briefly describe its action.

64. What is the meaning of the term 'amalgamation'? Name some useful amalgams, and state for what purposes they are used.

65. What metals can be transferred from one receptacle to another by distillation?

66. State what is understood by the term 'liquation' and name two metals which may be separated from each other by liquation.

67. Illustrate by an example in each case the meaning of the terms 'scorification' and 'cupellation.'

68. Under what circumstances would it be advisable to scorify a substance previously to cupellation?

69. When sulphides are roasted in ordinary reverberatory furnaces, but little of the sulphurous acid produced is utilized for the production of sulphuric acid. To what cause is this due and how may the difficulties be overcome?

70. What is a regulus or matt and what is a speise? How is each produced? Give examples.

71. What purpose is served by converting metals into regulus or speise?

72. Name some processes in which chlorine, both in the free and in the combined state, plays an important part.

73. What is understood by the terms 'dry' and 'wet' processes? Give an illustration of each.

74. Can you name any metals which are extracted from their ores by the agency of electricity, or any which are refined by the same means? If so, describe the principles of the methods adopted.

Slags and Silicates.

75. How are silicates generally classified?

76. What is meant by the expressions 'mono-silicate' and 'bi-silicate' as applied to a slag? Mention processes in which they would respectively be produced.

77. How are slags usually formed. What are their general characteristics?

78. A given slag is glassy in one case and crystalline in another; how do you suppose each has been produced?

79. When is a slag said to be clean?

80. Suppose a given slag to be infusible: what substances would be likely, when strongly heated with it, to render it fusible?

81. A slag consists of two parts—acid and base. Give an instance in which a body is acid in one slag and basic in another.

82. The gangue of some ores is self-fluxing. How is this possible?

83. What is the general rule with regard to the relative fusibilities of single and multiple silicates?

84. Suppose an ore to contain a sulphide as an impurity, how would you convert it into slag?

Refractory Materials.

85. What is fire-clay, and of what substances is it formed? How does fire-clay differ from sand?

86.* For what purposes would you use a material which has the following composition — Silica, 97; alumina, 1·5; lime, 1·5?

87. What materials containing much silica are used to resist high temperatures in metallurgical appliances?

88.* What is the composition of the materials you would employ for (1) lining a Bessemer converter; (2) for the hearth of a Siemens' regenerative furnace; (3) for the bed of a furnace used for cupelling argentiferous lead?

89. What materials are used in the composition of the various kinds of crucibles?

90.* What is the difference in composition and properties between Dinas bricks and Stourbridge fire-bricks?

91. What materials are used for making respectively brass founders' and steel founders' melting pots?

92.* Is fire-clay supposed to consist essentially of a substance of definite chemical constitution, and if so, what is its formula?

93. Mention cases in which slags are used as refractory linings for furnaces.

94. Is there any difference between graphite, plumbago and blacklead crucibles? and if so, explain the difference.

95. What are the essential properties of a good crucible?

96. How would you test a sample of fire-clay so as to ascertain its power of resisting corrosion? How would you test its refractory power?

97. What are the most injurious impurities in fire-clay?

98. Why is sand, graphite, or burnt clay added to raw clay in making crucibles? What would result if the raw clay were used alone?

99. Name any furnaces in which the beds are lined with sand.

100. How would you test a clay crucible in order to ascertain whether it is suitable for ordinary laboratory use?

APPENDIX. 181

101.* On analyzing two samples of clay, it is found that one of them contains 4 per cent. of potash and soda, while these substances are absent in the other. How would these clays behave at high temperatures?

102.* State the conditions under which you would employ (1) basic bricks, (2) acid bricks, in the construction or lining of metallurgical furnaces.

Fuel.

103. What is meant by the calorific power of fuel? State the calorific powers of hydrogen, wood charcoal and carbonic oxide.

104. What is meant by the terms 'perfect' and 'imperfect' as applied to the combustion of carbon and hydrogen?

105. State why any one kind of coal differs from another in heat-giving power.

106. What is the nature and composition of carbonic oxide, and how may it be produced from charcoal or coke?

107. What is meant by the term 'calorific intensity'? How is the calorific intensity of carbon and of hydrogen computed?

108.* Why is the temperature actually obtained in a furnace lower than that indicated by theoretical considerations based on the composition of the fuel employed?

109. Calculate the calorific power of CH_4 and C_2H_4.

110.* The calculated calorific powers of hydrocarbons do not agree with those determined by experiment. Account for this difference.

111.* How do you account for the fact that different forms of carbon, such as charcoal, graphite and the diamond, have different calorific powers?

112.* Describe any form of calorimeter with which you may be acquainted.

113. What is the composition of cellulose? What other substances are present in wood?

114. What substances are formed when wood is heated at a temperature above 200° in a closed vessel?

115.* What is the composition of the ash left when wood is burnt so as to remove all combustible matter?

116. What is the object of converting wood into charcoal? Mention some metallurgical processes in which charcoal is used.

117. Describe the method of making charcoal in circular piles and state what kinds of wood are most suitable for this purpose.

118. What is peat? How is it formed and for what purposes is it used?

119. What is the difference in composition between the ashes of peat and charcoal and how do you account for this difference?

120. What is coal? State how lignite, bituminous coal and anthracite differ from each other.

121. Name the constituents of the ashes of common coal.

122.* What proof is there that anthracite has been derived from vegetable matter?

123.* What indication is furnished by the proportion of oxygen in a fuel?

124.* How would you burn finely-divided coal so as to obtain the best results?

125.* Coal sometimes takes fire spontaneously on board ship. What is the reason, and how would you prevent it?

APPENDIX.

126.* For what metallurgical operations would you employ (1) a long flame non-caking coal, (2) a short flame caking coal?

127.* In the same coal basin are found both steam coal and gas coal. How do you explain this?

128. What is petroleum? How does it differ from coal?

129. What is the nature of the substance termed 'Boghead'?

130. What is 'cannel-coal' and for what purposes is it used?

131. What is peat charcoal and how is it made? Describe its properties.

132.* In what forms of combination does sulphur occur in coal? Would a sulphurous coal be as equally objectionable for use in a blast furnace as in a reverberatory furnace? Give the reasons for your answer.

133. How is common coke made? Describe the apparatus employed and the principles which should guide the coke burner.

134.* How would you proceed to examine coke in order to ascertain its suitableness for smelting iron?

135. Describe the construction of Cox's coke oven, and mention the advantages claimed for it over the ordinary oven.

136. Describe the construction and mode of working the Appolt coke oven.

137. Describe the Coppée coke oven and state what kind of coal is most suitable for conversion into coke by this method.

137A. What are the essential features of the Simon-Carvè coke oven? What useful products are obtained from it?

138. Sketch and describe any kind of kiln employed for making coke.

139.* Describe the properties and mode of manufacture of any kind of patent fuel.

140. What is the composition of the gases liberated from a common bee-hive coke oven ?

141. Name any methods of utilizing the waste gases of coke ovens.

142. Mention any methods for the desulphurization of coke, and explain the nature of the chemical reactions involved.

143.* Describe any coke oven having appliances for collecting the volatile products, and name the products thus utilized.

144. Sketch and describe any form of producer for making gaseous fuel.

145. What is meant by the terms 'air-gas' and 'water gas'?

146.* Describe the Wilson gas-producer, and state how the gases generated therein differ from the gases produced in Siemens' arrangement.

147. Show by a sketch a reverberatory furnace in which the ordinary fireplace is replaced by a gas-producer.

148. What difference in the atmosphere of a reverberatory furnace may be produced by burning (1) a thick layer of coal on the bars, (2) a thin layer ?

149. Explain why, in generating 'producer gas,' the presence of water in the air by which the fuel is burnt increases the amount of carbonic oxide, and state what is the limit of water vapour that would be beneficial.

150. What is the principle of the Siemens' open hearth furnace and for what purposes is this furnace used in metallurgical operations ?

151. Explain why certain furnaces are provided with dampers.

152. For what purposes is a chimney used in connection with a furnace ?

153.* What methods are employed for measuring high temperatures in furnaces ? Describe Siemens' electrical method.

153A. Describe any form of pyrometer with which you are acquainted and the principle on which its action is based.

154.* Upon what conditions does the economical working of a gas-producer depend ?

155.* What effects are produced by coal on the materials of the furnaces in which it is burnt ?

156.* Give a rough sketch of an ordinary fire, and describe the chemical condition of its various parts.

157.* State the effect of (1) a very slight, (2) a moderate, (3) a very powerful draught on the products of combustion of coal in a furnace.

158.* Show by sketches in plan and vertical section either Hoffman's circular kiln as used for burning firebricks, or Lürmann's kiln, used for the continuous coking of coal.

Iron.

159. What is iron ? How does it differ from cast-iron and steel and to what industrial uses is each specially applicable ?

160. State the physical properties of wrought iron.

161. What is understood by the term 'welding'? How does the property of welding depend on the physical condition of the iron ?

162. Why does a blacksmith use sand in welding iron ? Why is it more difficult to weld iron to steel than iron to iron ?

163.* State what precautions are necessary in welding steel and state why they are necessary.

164. What is meant by the tensile strength of iron and how is it determined ?

165.* State fully how you would test the mechanical properties of an iron bar.

166. What is the difference in structure between a bar of iron produced by rolling and a piece of the same bar after fusion and how may that difference be shown ?

167. State what occurs when bright iron is exposed to moist air and mention how iron may be protected from the action of the atmosphere.

168.* Show with the aid of a sketch the approximate dimensions of the portions of metal used in determining the tenacity of steel ; describe recent views as to the influence of the dimensions of the test piece on the result of the test.

168A.* What are the dimensions of a test bar of cast iron and how is its strength usually determined ?

169. What are the physical properties of pure iron ? State how you would prepare a small quantity of pure iron for purposes of research.

170. What is burnt iron, fibrous iron and crystalline iron ?   How is each produced ?

171. What is the effect of cold hammering on iron ?

172. Explain the terms 'hot short' and 'cold short.'

173.* What is the nature of the rust of iron, and what change, if any, would iron undergo if kept at ordinary atmospheric temperatures in pure water free from air ?

174. Name the three oxides of iron of metallurgical importance and state what you know of their respective uses.

175. What occurs when iron and a small quantity of sulphur are strongly heated together? Is there any difference when the sulphur is in excess, and if so, state the difference?

176. What occurs when sulphide of iron is roasted in air (1) at a low temperature, (2) at a high temperature?

177. What is the difference between ferrous sulphide and iron pyrites? State what you know concerning their respective uses.

178. When iron is treated with one variety of nitric acid it becomes 'passive.' What is this variety and how do you suppose it acts?

179.* What changes are produced in the appearance of the fracture of pig iron by the presence of sulphur in different proportions?

180. How may a rolled bar of good iron be broken so as to make it present either a fibrous or a crystalline fracture?

181. What occurs when iron is melted in contact with phosphate of lime, carbon and silica?

182.* What is the effect of different amounts of phosphorus on iron. In smelting phosphoric ores of iron, where would you expect to find the phosphorus? Give reasons for your answer.

183.* Iron containing phosphorus is strongly heated with carbon. What change, if any, takes place?

184. What is the effect of arsenic on iron?

185. Does any action occur when pure iron is strongly heated in admixture with silica, and if so, what is that action?

186. What is the effect of silicon on iron? What is the nature of the substance known as silicon-iron and for what purposes is it used?

187. How would you proceed in order to obtain a pig iron rich in silicon?

188. Forge and mill cinders consist largely of silicate of iron. Can you express their composition by formulae? and if so, write the formula in each case.

189. What is the name given to the substance left behind when tap cinder is submitted to liquation?

190. When tap cinder is smelted in the blast furnace, it is reduced by carbon; show by an equation the change which occurs.

191. Under what conditions will pure iron take up carbon and what is the maximum amount it will retain?

192.* Describe or show with the aid of curves the effect of the simultaneous presence of carbon and manganese on the tenacity of steel.

193.* What is the difference between grey, white and mottled pig iron? What is the effect of nitric acid on each variety?

194.* What is spiegel-eisen; how is it formed, and for what purpose is it employed?

195. What is the effect of different quantities of carbon on iron?

196. What is meant by the expressions 'hardening carbon,' 'free carbon,' 'graphite,' 'kish' and 'combined carbon'?

197. Manganese is very frequently present in iron; what influence does it exert on the iron? What is ferro-manganese?

198. A triple compound, of iron, silicon and manganese, is now manufactured. What is it used for? Explain its action.

199.* When a mixture of ferric oxide and silica is heated to strong redness with access of air, what takes place?

APPENDIX.

200. State what you know of the useful alloys of iron.

201.* Titanium is sometimes found in iron slags; what is the nature of the compound containing the titanium?

202. Describe the chief ores of iron.

203. Can iron be economically extracted from iron pyrites? If so, how would you prepare the ore for reduction?

204.* How would you analyze a sample of pig iron?

205.* How are iron ores assayed by the dry method? What fluxes would you employ respectively for haematite and clay ironstone?

Extraction of Iron.

206. How may iron be directly obtained in the malleable state from its ores?

207. How was iron extracted from its ores before the discovery of cast iron?

208.* Describe the Walloon process of making malleable iron.

209. Describe the extraction of iron in the Siemens' rotatory furnace, and state what advantages are claimed for the method.

210. Explain the way in which the earthy matters associated with iron ore are fluxed away in the Catalan process.

211. Point out how the chemical changes in the Catalan forge resemble the action in the puddling furnace.

212. Describe the arrangement for producing the blast used in the Catalan furnace.

213.* Describe the method introduced by Chenot and improved by Blair, for the direct extraction of iron, and state how far the method has been successful.

214.* When an ore consisting of ferrous carbonate is heated to strong redness in a vessel from which the air is excluded and the substance of which has no action on the ore, what chemical change occurs?

215. When iron is extracted from a phosphoric ore in a small blast furnace like the Catalan forge, what becomes of the phosphorus?

216. How would you smelt an iron ore so as to produce 'pig iron'? Give a sketch of the furnace you would use.

217.* What substances are generally used as fluxes in iron smelting? How do they act and for what kinds of ores is each most suitable?

218. What do you understand by the term 'weathering' as applied to iron ores and what useful purpose does it serve?

219. For what purposes are iron ores calcined and for what ores is the process unnecessary?

220. Describe a modern form of kiln for calcining iron ores.

221.* The earthy matter in some iron ores is self fluxing, what would probably be the composition of such an ore?

222.* Iron obtained by reducing forge and mill cinders is very impure, mention any methods which have been devised for obtaining a purer product.

223. Iron ore, fuel and flux are charged into the top of a blast furnace, and pig iron is tapped out at the base of the furnace. Explain how the carbon acts on the ore so as to render this result possible.

224. Explain how it is that iron obtained from the blast furnace is impure. How much carbon is found in ordinary grey pig iron?

APPENDIX.

224A. How do grey, white and mottled pig iron differ from each other with regard to the condition of the carbon?

225. What is the nature of the gases collected from the top of a blast furnace and for what purposes are they used?

226. What are the advantages and disadvantages of the hot blast in iron smelting?

227. The mode of charging has a great influence on the proper working of a blast furnace for smelting iron. Describe a good mode of charging a large furnace.

228.* Describe three modes of closing the top of a blast furnace, and state the relative merits of each.

229. Give a sketch of the appliance known as the cup and cone, used in charging a blast furnace.

230. State clearly what conditions in a blast furnace tend to the production of impure varieties of pig iron. What are the usual impurities?

231. The waste gases from a blast furnace used for smelting-iron are combustible: account for this.

232.* Distinguish between the nature of the charge and the temperature and pressure of the blast when iron is smelted with a view to the production of 'white' or 'grey' pig iron respectively.

233.* Explain why in a blast furnace of large dimensions the distance between the twyers, measured across the hearth, must be carefully considered and adjusted.

234.* What are the constituents of the slag which accompanies the production of pig iron from the clay ironstone of the coal measures, and what on inspecting such a slag would lead you to infer that it probably contains phosphoric acid in sensible quantity?

235. Draw a vertical section of a modern blast furnace and name the various parts.

236.* State the reasons that may be assigned for the inner form and large dimensions of the modern blast furnaces used in this country for iron smelting.

237.* Give the formula for the crystallized slags produced in iron smelting.

238.* Show by a diagrammatic sketch the various zones in a blast furnace, and write the probable temperatures in the different zones.

239.* What is the composition of the slag produced (1) in the blast furnace during the smelting of iron ores; (2) during the smelting of manganiferous ores for the production of spiegel-eisen?

240.* Describe minutely the appearances of the fracture of different kinds of pig iron, and state the conditions in the process of smelting upon which those fractures depend.

241. What indications are afforded by the character of the slag from an iron smelting blast furnace with respect to the working conditions of the furnace and the quality of the pig iron produced?

242.* What reasons have been assigned for the deep blue coloration of the slags which are occasionally produced in iron smelting furnaces.

243. Describe the Rachette furnace, and state what reasons are assigned for its particular mode of construction.

244.* Describe the first form of apparatus employed for heating the blast for an iron smelting furnace, and state what were the chief defects of such an arrangement.

245.* What modifications were adopted to remedy the defects referred to in the last question?

246. Give a sketch in sectional elevation of a Whitwell stove for heating the blast.

APPENDIX.

247. Give a sketch in vertical section of an ordinary 'pipe stove' used for heating a blast of air.

248. Give a sketch in sectional elevation of a Cowper stove for heating the blast, and state how it differs from Whitwell's.

249. The blast is sometimes heated on the principle of conduction and sometimes on the regenerative principle: explain these terms.

250. What is the function of the combustion chamber in a Cowper or a Whitwell stove?

251. What are the chief chemical reactions which take place in the production of pig iron?

252.* How is ferro-manganese made on the large scale, and for what purposes is it used? State its physical properties and approximate chemical composition.

253. If you were required to produce pig iron, 'rich' and 'poor' in silicon respectively, how would you proceed?

254. Why should less fuel be used in smelting with hot blast than in smelting with cold blast?

255.* In smelting iron ores which contain lead, zinc, titanium, and chromium respectively, what becomes of those metals?

256.* Sketch and describe a modern form of hot blast water twyer.

257. Different methods are employed for collecting the waste gases of a blast furnace, mention as many as you know of.

258.* How is spiegel-eisen prepared, and for what purposes is it used? What is its composition, and what are its physical properties?

259.* Mention the various uses of the cast iron obtained from a blast furnace, and state the various forms into which it is tapped.

260. What is the meaning of the terms 'scaffolding,' 'bears' and 'slips' as applied to a blast furnace.

261.* What is the average daily make of an old charcoal and of a modern coke blast furnace for smelting iron?

262. How do you account for the fact that pig iron produced with coke as fuel and with hot blast, is of different character to pig iron produced by the aid of charcoal and with cold blast?

Production of Malleable Iron from Cast Iron.

263. What is the difference between malleable iron and malleable cast iron and how is the latter produced?

264. Wherein does the old method of producing malleable iron from pig iron differ from the modern method?

265. State the principles of the process by which cast iron may be converted into malleable iron.

266. Describe the process of refining pig iron and state why that process is now seldom used.

267. Wherein does the modern differ from the original process of puddling?

268. In what way does 'dry puddling' differ from the 'pig boiling' process?

269.* What modifications of Cort's process of puddling have been proposed from time to time?

270. Why is 'white' pig iron more suitable for the dry puddling process than 'grey' pig iron?

271. The presence of manganese in pig iron is considered advantageous; what are the reasons for this?

272. State fully the effect of 'chill casting' on grey pig iron.

APPENDIX.

273. Describe what takes place when pig iron is strongly heated in a 'reducing' atmosphere in the presence of silica.

274. Describe a process in which pig iron is converted into steel in an ordinary reverberatory furnace.

275. How may malleable iron be obtained comparatively free from phosphorus from an ore containing a large amount of that element?

276.* What methods have been tried for conducting the operation of puddling with mechanical tools?

277.* What are physical and chemical properties of a 'puddle-ball'? What is meant by the term puddling?

278. What becomes of the silicon, manganese, phosphorus, sulphur and carbon during the conversion of pig iron into malleable iron in the puddling furnace?

279. What is tap-cinder? How is it prepared for fettling the bottoms of puddling furnaces? What is the reason for such preparation?

280.* How has the waste heat of puddling furnaces been utilized? Describe any arrangement you know of.

281.* What are the physical and chemical properties of 'bull-dog' and 'bull-dog slag'?

282. What modifications of the puddling process are necessary when producing steel instead of iron?

283.* Mention the different kinds of fettling that have been used for puddling furnaces.

284. Give a sketch in sectional elevation of an ordinary puddling furnace?

285. Give a sketch in sectional elevation of a refinery?

286.* Give sectional drawings of a Danks' puddling furnace, and state what advantages are claimed for the method

287.* Describe any form of gas furnace that has been used for puddling, and state how far it has been successful.

288. What do you understand by the expression 'mechanical treatment of the puddle-ball'?

289.* Enumerate the various appliances used for shingling puddled iron.

290.* A forge train contains two sets of rolls; what is their nature, and what is the function of each set?

291.* What is meant by the terms 'puddled bars,' 'slit rods,' and 'plate rolling'?

292. How is the iron obtained which is used for the manufacture of what are termed 'charcoal plates'?

293.* What is the difference between the manufacture of 'charcoal plates' and that of 'coke-plates.'

294.* In what respects does the Lowmoor process of manufacturing iron differ from the ordinary process?

295. Describe the 'open fire' or 'hearth finery,' and state for what purpose it is used.

296. Give a sectional drawing of the re-heating or mill furnace and state for what purpose it is used.

297. What is the difference between the mill furnace and the puddling furnace?

298. What is the nature of the slag from the re-heating furnace, and for what purposes is it used?

299. What is meant by the terms 'piling' and 'fagoting' as applied to iron? What are merchant bars?

300. What purpose is served by re-heating puddled ron, and how far is the method successful?

301. What kind of lining is used for the mill furnace, and what is its action on the iron heated upon it?

302.* Describe the nature and action of any very recent form of lining for re-heating furnaces.

303.* Describe the Ponsard re-heating furnace and explain the principle of the heat recuperator.

304.* What is the essential difference between the Siemens and Ponsard principle of heating?

305.* Explain the Boetius method for heating mill furnaces.

306.* Explain the principle of Bicheroux' gas furnace.

307.* What is the nature of Russian sheet iron; how is it made and to what purposes is it specially applied?

Steel.

308. What is steel? How does it differ from malleable iron and cast iron? Why is steel so valuable in the arts?

309. Enumerate the different processes employed for making steel.

310. Describe the processes of hardening and of tempering steel.

311. Is the bulk of a piece of steel changed by the process of hardening, and if so, in what manner?

312. What amounts of carbon are generally accepted as constituting the difference between malleable iron, cast iron and steel?

313.* What constitutes the difficulty of soundly welding iron and steel together?

314.* In what states does carbon exist in steel hardened in the usual manner, and in the same steel after annealing?

315.* What are the physical properties and chemical composition of the steel now used for ship plates? State why such plates are preferred to plates of wrought iron.

316. What is blister steel? Describe its properties and enumerate the different modes of treating it so as to produce commercial steel.

317. What is shear steel, and how is it made?

318. How may malleable iron be converted into steel?

319.* Describe the principle of Euchatius' method of converting cast iron into steel.

320. What analogy is there between the production of blister steel and the case hardening of iron?

321. Give a sectional drawing of a cementation furnace, and describe the uses of the various parts.

322. How do you explain the conversion of iron into steel in the cementation furnace?

323. What is the difference in the character of a bar of iron before and after conversion into steel? How is the completion determined?

324. If air has gained access to the bars during conversion into steel, what is the effect produced?

325. How is iron case hardened?

326.* What principles should guide you in the selection of bar iron with a view to its conversion into steel?

327. How is 'crucible steel' made, and to what purposes is it applied?

328.* What precautions are necessary in casting steel after melting in crucibles, so as to obtain sound ingots?

329. Describe Huntsman's and Mushet's methods respectively, of making steel.

330.* Describe by the aid of a sketch, Siemens' regenerative furnace used for melting steel in crucibles.

331.* Describe Heath's manganese process for making steel. How do you explain the influence of the manganese?

332. Describe one method of converting pig iron into steel.

333. What is puddled steel, and how is it made?

334.* Describe the Heaton process of making steel?

335.* With what results has the use of fluor-spar been employed in making steel, and how do you explain its action?

336.* How did Mr. Parry propose to make steel of high-class quality on the large scale?

337. How is steel made in the Catalan and similar forges?

338. Describe the acid Bessemer process of making steel, and state the composition of the converter lining.

339. Describe the construction of the Bessemer converter, and illustrate your answer by sketches.

340. If you were asked to report on a sample of iron ore with regard to its suitableness for producing pig iron fit for the Bessemer process, by what principles would you be guided?

341. Why is the presence of much silicon considered advantageous in pig iron intended for conversion into steel by the Bessemer process?

342. Suppose pig iron containing silicon, phosphorus and sulphur, to be treated in the ordinary Bessemer process, what becomes of these elements?

343. What is Spiegel-eisen and for what purpose is it used in the Bessemer process?

344. How would you propose to work up the crop ends of rails or other steel scraps?

345. Why does the bath of metal in a Bessemer converter remain fluid when air is blown through it?

346.* Ingot steel produced by the Bessemer process contains blowholes. In what portion of the ingot are these most abundant? How do you account for them, and how may they be avoided?

347.* Show by means of a sketch, in vertical section, the arrangement of a hydraulic crane and ladle used in connection with the Bessemer process.

348. What is meant by the 'basic' Bessemer process, and how does it differ from the acid process?

349.* What amount of phosphorus does Bessemer basic slag contain? How has such slag been utilized?

350.* Indicate by means of a curve the order in which the constituents of the molten metal are attacked during the 'blow' in either an acid or a basic Bessemer converter.

351.* It has been observed that at certain stages of the basic Bessemer process the iron contains calcium and magnesium. How do you account for this, and how is it that these elements are not present in the finished iron?

352.* Show, by the aid of a sketch, the arrangement of the Clapp-Griffith modification of the Bessemer converter. State what advantages are claimed for the process, and describe the character of the metal produced.

353. Describe the method of making steel in the Siemens' regenerative furnace.

354.* Describe, with the aid of sketches, the arrangements recently proposed for distributing the gas and air in a Siemens' regenerative furnace.

355.* Explain the nature of the changes recently proposed by Siemens in the construction of the regenerative furnace.

APPENDIX.

356.* The regenerators in a Siemens' steel furnace have recently been built in a different position than formerly. Explain the new arrangement, and state what advantages are gained by it.

357. What is meant by the term 'mild steel,' how is the metal produced on the large scale, and to what purposes is it specially applicable?

358.* Why is steel cast under great pressure, as in Whitworth's process, alleged to be superior to the same kind of steel cast under the ordinary pressure?

359. What is meant by 'Siemens' steel and 'Siemens-Martin' steel, and what is the distinction between them?

360.* Describe the recent method of making steel in a Siemens' furnace with a basic lining. What is the nature of this lining and for what purposes is it used?

361.* Write a short account of the methods adopted for the elimination of phosphorus from iron employed for making steel.

362. Why is steel prepared by the Bessemer and Siemens processes, usually not well suited for the manufacture of the best kinds of cutting instruments? What class of steel is preferred for this purpose?

363.* A watch spring requires to be made of steel, having great toughness and elasticity. How is such steel made, and how does it differ from the steel required for a surgical knife?

364.* Can you suggest any process by which phosphoric acid may be profitably extracted from the basic Bessemer slag?

365.* What is the composition of the gases escaping from the mouth of the Bessemer converter?

SILVER.

366. State the physical properties of silver.

367. What is meant by 'spitting' of silver, and how do you explain the phenomenon?

368. What occurs when molten silver solidifies rapidly, and when the metal is allowed to cool slowly out of contact with air?

369. Silver is said to be a soft metal; how does it compare with copper and gold in this respect?

370.* What is the melting point of silver? Describe a method of determining this point.

371. What action occurs when silver is heated with sulphur? Express the change by an equation.

372. State the physical properties of silver sulphide.

373. What occurs when silver sulphide is heated with iron, lead and copper respectively?

374. What occurs when silver sulphide is roasted (1) alone, (2) in admixture with iron and copper sulphides?

375. Describe by an equation what occurs when cupric chloride is heated with sulphide of silver.

376. What is understood by the term 'oxidized' silver and how is it produced?

377. When silver articles are exposed to the air of large towns they become tarnished. Explain this.

378. Describe any good method of removing the tarnish from silver goods.

379. What is the composition of the liquid employed for stripping the silver from electro-plated goods?

380. How is sulphate of silver produced on the large scale?

381. What is the formula for silver hyposulphite, and how may this salt be dissolved?

382. How is silver chloride formed? What occurs when silver chloride is strongly heated for some time in a clay crucible?

APPENDIX. 203

383. Name several liquids which dissolve silver chloride, and show by equations the changes which take place.

384. How is silver chloride formed from silver sulphide on the large scale (1) in a furnace, (2) in the open air?

385. What substances are usually employed for reducing silver sulphide? Show the re-actions by equations.

386.* Pure carbon will not reduce silver chloride; how then do you account for the fact that silver chloride is reduced by charcoal or coke?

387. Give the composition of some useful alloys of silver.

388. What are the most common impurities in commercial silver?

389. What are silver solders? Give the composition of those known to you.

METHODS OF EXTRACTING SILVER FROM ITS ORES.

390.* What is the principal source of silver in the United Kingdom?

391. Enumerate the chief ores of silver.

392. What is the nature of metalliferous minerals containing silver other than silver ores proper?

393. Describe briefly the processes employed on the large scale for extracting silver from its sulphuretted ores by the dry way.

394. What re-actions occur in the Mexican amalgamation process for silver ores?

395. Describe the Mexican amalgamation process for extracting silver.

396. Describe a process by which silver may be separated from rich copper regulus by the dry way.

397. Describe a process for extracting silver from rich copper regulus by the wet way.

398. Explain, by the aid of a sketch, how you would separate silver from mercury.

399.* How would you proceed to treat an ore containing native silver and chloride of silver so as to extract the silver?

400.* In the Claudet process for the desilverization of roasted copper pyrites the silver is precipitated as iodide. How is it that this process is commercially successful, *cuprous* iodide also being insoluble?

401. How is silver separated in the wet way from a solution which mainly consists of chloride of copper?

402.* In smelting silver ores containing much barium sulphate, with lead, what modifications in the blast furnace treatment are necessary?

403. How is silver extracted from burnt copper pyrites?

404.* Show, by a sketch, the essential details of any form of 'pan' used for amalgamating silver ores.

405. What is the difference between the Augustin and Ziervogel processes of extracting silver?

406. Describe the barrel process for the extraction of silver.

407. How are silver and gold extracted from mixed ores at Freiberg?

408.* Explain the process devised by Von Patera for extracting silver.

409.* Describe the old methods of silver plating.

410. How would you proceed to electro-plate a brass spoon with silver? State how you would make the solution and what source of electrical power you would employ.

APPENDIX. 205

411. What are the essential points in the extraction of silver by the Ziervogel process ?

412.* Describe the method of assaying silver alloys by cupellation.

413.* An ore contains both the sulphide and the sulph-arsenide of silver. How would you treat it ?

414. A silver ore has been roasted with salt, and the silver in the roasted material is present as chloride. How would you extract the chloride and precipitate the silver in the metallic state ?

GOLD.

415. Enumerate the chief physical properties of gold.

416.* State what you know concerning the causes of the occasional brittleness of the gold used for coinage in this country.

417.* What methods are employed to deprive brittle gold of its brittleness, and what are the principles of those methods ?

418. When gold is maintained at a white heat for a considerable time, what occurs ?

419. Very small quantities of foreign substances in gold tend to make it brittle. State what you know of this subject.

420.* A stream of oxygen, sulphur vapour, and chlorine respectively, is passed through molten gold ; state what occurs in each case.

421.* Hydrochloric and nitric acids have no action separately on gold, but when mixed, the gold is dissolved. How do you account for this ?

422.* Gold is precipitated from its solutions by most metals. How do you explain this fact ?

423. What are properties of gold which make it so useful as a circulating medium?

424.* Commercial gold is never pure. What are the most frequent impurities?

425. How would you prepare a sample of pure gold?

426.* Gold containing small quantities of lead, antimony and tin is too brittle to roll; state how you would remove these impurities.

427. What is meant by the expressions '18 carat' and '22 carat' as applied to gold?

428.* There is a rough method of testing the quality of gold by means of the touchstone. Explain this method.

429.* Describe the process of gold beating. How is it that gold can be reduced to such an extreme degree of thinness?

430.* What is the composition of the peculiar compound termed 'purple of cassius,' and how is it prepared?

431. Name some useful alloys of gold used in jewellery.

432. What is the general effect of copper, silver, tin, zinc and lead respectively upon gold?

433. If you wished to prepare alloys of gold having (1) light yellow, (2) greenish, (3) red tints, how would you proceed?

434. Give the composition of some cheap alloys of gold.

435. How does gold occur in nature and with what metals is it generally associated?

436. Explain the terms 'auriferous quartz,' 'auriferous pyrites,' and 'alluvial deposits' of gold.

437.* A sample of gold is found to contain platinum and iridium. How would you separate these metals from the gold?

438. What would you look for when analyzing native gold ?

439. How do you account for the fact that gold is generally found in nature in the metallic state ?

Methods of Extracting Gold.

440. How may gold be extracted on the large scale from auriferous pyrites ?

441. How would you extract, by a furnace method, the precious metal from quartz containing gold ?

442. If you had gold in solution as chloride, how, on a large scale, would you precipitate the gold ?

443.* When, during the smelting for copper, gold ores are added to sulphide of copper, in what products would you expect to find the precious metal ?

444.* When a copper regulus containing gold and silver is desilverized by the Ziervogel process, what becomes of the gold ? State how you would extract the gold.

445.* The 'tailings' from pan amalgamators are found to be rich in the precious metals. How would you treat these 'tailings' so as to recover the gold and silver ?

446.* How would you extract gold and antimony from an auriferous stibnite ?

447. Describe the Plattner process for extracting gold.

448. Describe the old method of extracting gold by the amalgamation of gold quartz.

449. Describe a recent method of extracting gold by amalgamation of gold quartz, without the aid of heat or water.

450. Under what circumstances would you extract gold from an ore by the aid of (1) lead and (2) mercury ?

451. Mercury used in the amalgamation of gold ore is sometimes said to be 'sick,' explain this expression.

452. Name any recent electrical method devised to prevent the sickening of mercury used for amalgamation of gold.

453.* What is meant by the term 'pan amalgamation'? Describe any form of pan used for this purpose.

454. How is gold amalgam treated so as to separate the gold ?

455.* Describe any process for the extraction of gold from quartz or pyrites by a smelting operation.

456.* What methods have been devised for extracting gold from its ores by means of lead and iron?

457.* Describe any form of revolving furnace in which gold ores are roasted with common salt.

458. A chlorine method of extracting gold from its ores has been devised in which the chlorine is produced by a current of electricity. Describe this method.

459. What is the action of chlorine gas on molten gold and on molten silver respectively ?

460.* State the principles of the process employed by refiners to recover the precious metals from jewellers' sweep and other shop waste.

461.* What are the relative advantages and disadvantages of Cornish rolls and of stamps employed in the crushing of auriferous quartz ?

462.* Metallic bismuth is found to contain gold. How would you extract the gold ?

Separation of Gold from Silver and Copper.

463. An alloy contains 20 per cent. of gold and 80 per cent. of silver. How would you separate the two metals ?

APPENDIX.

464.* Describe the method of parting by sulphuric acid, and express the re-actions by equations.

465. Describe the nitric acid method of parting, and show how it differs from the sulphuric acid process.

466. Describe a dry method of parting, and state under what circumstances the wet method is preferable.

467.* Describe an arrangement for condensing the acid fumes given off during the operation of parting.

468. State what indications are afforded by the chlorine method of parting as to the completion of the operation.

469. Describe a dry method of separating gold from copper.

470. What is the action of chlorine gas on molten gold containing silver?

PLATINUM.

471. What kind of a metal is platinum, and for what purposes is it used in the arts?

472.* What is spongy platinum or platinum black, and how is it prepared?

473.* What use is made of spongy platinum in the arts?

474. How does platinum occur in nature? What is the chief source of the metal?

475.* What metals and minerals are often found associated with platinum in nature?

476. Describe Deville's method of separating platinum from the metals with which it is generally associated.

477. Describe the method of melting platinum on the large scale.

478. Describe the wet method of extracting platinum.

479. What is meant by 'platinizing' and 'platinating,' and how is each process carried out?

o

Copper.

480. What are the chief physical properties of metallic copper?

481.* At what temperatures is copper malleable?

482. Name the two oxides of copper, and show how they differ from each other.

483. Pure copper is known to possess high conductivity; explain fully the meaning of this.

484. When copper is heated to redness in air what is the composition of the scale?

485. Mention the most common impurities of commercial copper, and state which are the most injurious.

486. Name the varieties of commercial copper, and state which is the most pure. Give reasons for your answer.

487. State the composition and properties of the compound formed when copper is heated in a crucible with excess of sulphur.

488. When a mixture consisting of 10 parts by weight of red oxide of copper and 7 of cuprous sulphide is heated to redness in a covered crucible, what is the composition of the product?

489. State what occurs when sulphide of copper is strongly heated in a current of steam. Express the change by an equation.

490. What occurs when sulphide of copper is roasted (1) at a low temperature, (2) at a strong red heat.

491. How is the regulus known as white metal converted into metallic copper?

492. If a small quantity of arsenic is present in copper, how will the physical properties of the copper be affected?

APPENDIX.

493. How is it possible to reduce sulphide of copper by the action of heat and air alone ? Explain with the aid of equations.

494. What takes place when a mixture of silica and black oxide of copper is exposed to a high temperature ? Express the change, if any, by an equation.

495. What occurs when cuprous oxide is strongly heated with sulphide of iron and silica ?

496. What occurs when cuprous sulphide is heated (1) out of contact with air, (2) with access of air ?

497. What is the nature of the green substance formed when copper is exposed to air and moisture ?

498. What change is produced in the appearance of the fracture of commercial copper by the presence of cuprous oxide ?

499. Describe the effect respectively, of carbon, silicon, arsenic, antimony and iron, when strongly heated with pure copper.

500.* How would you prepare a small sample of pure copper ? Mention any means of distinguishing it from ordinary commercial copper without the aid of chemical analysis.

501. Explain the terms 'overpoled,' 'underpoled,' and 'tough-pitch' copper.

502. What is the effect of heating sulphide of copper with lead and iron respectively ?

503.* What is the effect of heating impure copper with nitre ?

504. What is the effect of phosphorus on copper ?

505.* When oxide of copper is strongly heated in a clay crucible, the latter is strongly corroded. Explain this action, and mention the compounds likely to be formed.

506.* What change occurs when cupric oxide is strongly heated with sulphide of lead ?

507.* Describe the copper ores of Cornwall and Devon.

508. Enumerate the chief ores of copper, giving the chemical formulae and the possible percentage of copper in each.

509. Describe the nature and properties of the chief British copper ore.

510. What is the composition of copper pyrites and iron pyrites, and how can one be distinguished from the other without the aid of chemical analysis ?

511. If copper pyrites be roasted in air, what change occurs ?

512. Explain the terms 'oxidized' ore, 'yellow' ore and 'grey' ore, as applied to the ores of copper.

513.* Describe the Cornish method of assaying copper ores.

514.* In the dry assay of copper ores, why are some ores first converted into regulus and slag ?

515.* What is the action of nitre, borax, glass, lime, and fluor-spar respectively, in copper assaying ?

516.* Why is 'mundic' added to some copper ores for the purposes of the dry assay ?

517.* What is refining flux, and for what purpose is it used in copper assaying ? What is the nature of the slag produced from refining copper ?

518.* For what purposes are tartar, common salt, and carbonate of soda used in copper assaying ?

519.* Describe any wet method for assaying copper ores.

520.* Describe the coloration test for copper slags.

521.* If asked to make a chemical examination of commercial copper, how would you proceed ? What impurities would you expect to find ?

522.* Describe an electrical method of assaying copper and sketch the apparatus you would employ.

523.* Impure lead and pure copper are separately melted, and the molten metals are each allowed to stand for some time in an oxidizing atmosphere. How would this treatment affect the properties of the metals after they have solidified ?

COPPER SMELTING.

524. Arrange in a tabular or diagrammatic form a scheme showing the process of copper smelting as conducted in South Wales, giving roughly the percentage composition of the product in each case.

525.* State and classify the principles of the various processes which have been employed in smelting copper.

526.* In smelting copper as practised in Lancashire, what becomes of the arsenic, antimony, tin, nickel, cobalt, gold and silver ?

527.* Some of the older heaps of copper slag at Swansea are known to contain nearly 2 per cent. of copper. In what way would you treat them so as to extract the copper, the ordinary Welsh method not being applicable ?

528. What use is made of oxidized ores in the Welsh process of copper smelting ?

529.* What methods have been used for collecting and utilizing the sulphurous acid evolved during the roasting of copper ores ?

530. Sketch and describe the Gerstenhofer calciner.

531. What is the composition of the gases evolved from an ordinary copper calciner ?

532. What are the nature and composition of 'coarse metal' and ore furnace slag ?

533. How would you convert 'coarse metal' into 'white metal'? Describe the furnaces employed.

534. Describe the nature and properties of the substance termed 'blue metal.'

535. How is the occurrence of metallic copper in blue metal to be explained?

536.* What is tile copper and how is it produced in the Welsh process of copper smelting?

537. Give the prominent characteristics of pimple metal, and describe the mode of its formation.

538.* What occurs when sulphur is heated respectively with 'coarse metal' and 'blue metal'?

539. What occurs when metallic iron is strongly heated with silicate of copper?

540.* What are the nature and properties of the substance termed 'moss copper'? How do you account for its formation?

541. Describe the method of producing 'blister copper' in a reverberatory furnace. Mention its properties.

542. What are the substances known as 'clean' slag, 'sharp' or metal slag, and 'roaster' slag. What amount of copper would you expect to find in each?

543.* State what substances have in recent years been added to the bath of copper during the operation of 'poling,' and explain their action.

544. Describe the Welsh method of refining copper, and state the chemical changes which probably take place.

545. State the physical properties of 'dry' copper and the cause of those properties.

546.* Describe the process of converting 'blister copper' into 'tough cake copper,' and state the accompanying chemical re-actions.

APPENDIX.

547.* Describe the method of converting rich copper regulus into 'blister copper' by the Welsh method, and state what re-actions take place.

548.* Describe the process of making the variety of copper termed in commerce 'best selected,' and state the principles of that process.

549.* Napier proposed certain improvements in copper smelting; discuss the advantages and disadvantages of his method.

550.* Describe Rivot and Phillips' method of copper smelting, and state your opinion of the process.

551.* How is copper extracted from its ores in Japan?

552. Give an outline of copper smelting as carried out in Sweden and describe the furnaces employed.

553. Describe the method of refining black copper in Sweden, and state the accompanying re-actions.

554.* Describe the method of smelting copper schist in Prussian Saxony.

555.* What is the nature of the Mansfeld copper schist, and what metals are economically extracted from it?

556.* Describe the Mansfeld blast furnace for smelting copper and state the nature of the products obtained therefrom.

557.* At Perm, in Russia, a cupriferous pig iron is obtained. State the nature of this substance and mention the other substances with which it is accompanied.

558.* How does the Welsh method of treating copper ores in reverberatory furnaces differ in principle from the blast furnace method?

559.* How would you obtain a fusible slag when smelting in the blast furnace, lead and copper ores rich in zinc?

560.* Describe the method recently adopted for extracting copper from copper regulus by the Bessemer process.

561.* The copper ores of the Mediterranean coasts are frequently accompanied by considerable quantities of arsenical compounds. How would you treat such ores?

562.* How would you smelt a copper ore mixed with 25 per cent. of galena, with a view to extract both the lead and the copper?

563. What is meant by the term 'kernel roasting'? State how the process is conducted.

564. Make a drawing of the kiln employed at Agordo for roasting copper ores.

565. What is the nature of the 'kernel' obtained in kernel roasting, and what is the composition of the outer shell?

566. In kernel roasting, what bye-products are obtained and how are they collected?

567.* What is the theory of the process of kernel roasting, and how does it resemble the Welsh method of copper smelting?

568. Describe some simple wet method of extracting copper, and give the chemical re-actions which take place.

569.* Describe Bankart's method of extracting copper, and state how far it has been successful.

570.* How did Escalle propose to extract copper? Explain with the aid of chemical equations.

571. Describe the principles of the Henderson process for extracting copper.

572. What is 'cement copper'; how is it obtained, and what is its composition?

573. How is the copper precipitate, obtained by the Henderson process, purified so as to obtain commercial copper?

APPENDIX.

574. Give an outline of the method of extracting copper as conducted by the Tharsis Company.

575. Some copper ores treated by the wet method contain gold and silver. How are these metals separated from the copper?

576. How is copper ore, after roasting with salt, dissolved, and what is the nature of the liquid employed?

577.* Describe any form of furnace in use at the present time for roasting copper ores with salt.

578.* Describe the Hunt and Douglas method of extracting copper, and give the chemical re-actions which take place at each stage of the process.

579.* Give an account of the present state of the electro-metallurgy of copper.

580. In what way may copper be deposited on a large scale from solutions of chloride of copper?

581. If you wished to coat iron with copper, how would you proceed, and what solution would you employ?

582. Electro deposited copper is said to be remarkably pure. How do you account for this?

583. Suppose an impure copper anode to contain zinc, iron, tin, lead, nickel, and antimony; in what order would these metals be dissolved, and in what order, if possible, would they be deposited on the cathode?

584. Define the terms anode, cathode and electrolyte as applied to an electro depositing bath.

585.* What are the chief difficulties in the way of successfully refining copper by electricity?

586.* In refining copper, lead is sometimes added to the molten bath; why is this done, and under what circumstances should it not be added?

587.* Two slags have to be treated, one containing lead, and the other copper; what methods would you employ in order to extract these metals?

Alloys of Copper.

588. Give the composition of four of the principal alloys of copper, and state for what purposes they are suitable.

589.* Give the prominent characteristics and physical properties of the alloys of copper and tin.

590.* What are the nature and properties of phosphor bronze, silicon bronze, aluminium bronze and manganese bronze ?

591. Name the chief alloys of copper and zinc ; state their composition and general properties.

592.* Mention any useful alloys of copper containing iron, and state the general effect of iron on copper.

593. How was brass made in ancient times ? What is the difference between this and the modern method ?

594.* Are there any reasons to suppose that some alloys of copper and zinc are true chemical compounds, and if so, what are those reasons ?

595.* What are the nature and properties of Muntz' metal, and of what metals is it now composed ?

596.* Describe the appearance of the fractured surfaces of gun metal, bell metal, speculum metal and yellow brass.

597.* Describe an alloy of copper containing arsenic, and state for what purpose it is used.

598. What precautions are necessary in making brass ? Give reasons for your answer. Why is lead sometimes added ?

Zinc.

599. What are the physical properties of pure zinc, and how does it differ from the commercial metal ?

600. What is the effect of air, and of air and moisture, on zinc at the ordinary temperature?

601. Spelter is termed a brittle metal, how then can it be rolled into sheets?

602. Zinc is hardened by rolling, how can it be prevented from cracking during the process?

603. In what way would the rolling of zinc affect its specific gravity?

604. What substances would you look for in analyzing commercial zinc, and what indication as to the purity of the metal is afforded by its fractured surface?

605. When zinc is strongly heated in contact with air what occurs? What is the difference if the air be excluded?

606. Mention the chief practical applications of zinc in the arts.

607. Zinc is often contaminated with iron; how can the iron be removed?

608. How is lead separated from zinc? What is maximum amount of lead present in commercial zinc?

609.* Do zinc and phosphorus, or zinc and arsenic combine when strongly heated, and if so, what is the nature of the bodies formed?

610. How can zinc and sulphur be made to unite, and what is the nature of the product?

611. What occurs when zinc sulphide is roasted with free access of air?

612.* State the nature of the chemical change which occurs when sulphide of zinc is fused with nitre.

613.* How is silicate of zinc produced, and how is the zinc separated from it?

614.* Silicate of zinc is only fusible with difficulty, what substance would you add so as to render it readily fusible?

615.* State the nature and properties of the substance known as 'zinc-white,' and state for what purposes it is used.

616. What occurs when zinc oxide is strongly heated respectively, with iron, carbon, and sulphur?

617. What takes place when two equivalents of oxide of zinc and one of sulphide of zinc are strongly heated together in a closed vessel?

618.* Describe the following substances—Blende, calamine, red zinc ore and electric calamine; state their composition and the purposes for which they are employed.

Methods of Extracting Zinc.

619. How is zinc extracted on the large scale from calamine and from electric calamine?

620.* What are the advantages and disadvantages of the English, Belgian and Silesian methods of extracting zinc?

621. Describe the construction and mode of working the Silesian zinc furnace.

622. What re-actions occur in the extraction of zinc from calamine and from blende?

623. Describe the Belgian process of extracting zinc from its ores.

624.* Show by a sketch the way in which gaseous fuel may be used for heating any form of furnace employed for the extraction of zinc from its ores.

625.* Show by a sketch in vertical section the arrangement of a Silesian zinc furnace with regenerative chambers.

626.* Describe with the aid of a sketch a Belgian furnace for extracting zinc, arranged so that the waste heat passes to beds on which the ore is roasted.

APPENDIX.

627.* What difficulties have hitherto prevented the successful extraction of zinc from its ores in blast furnaces?

628. How is the crude zinc as obtained from its ores, purified, so as to prepare it for commerce?

629.* Describe the old Carinthian method of extracting zinc.

630.* What is the nature of zinc fume, and how is it treated in order to recover the zinc?

631.* Describe the Montefiori furnace, and state for what purpose it is used.

632.* What are the difficulties in the way of the electrolytic treatment of zinc ores? Describe such a method for the separation of zinc from zinc blende.

633.* Are there any wet methods for extracting zinc? If so, describe one of them; if not, what do you suppose to be the reasons?

634. What occurs when zinc and lead are melted together and allowed to cool slowly?

635. Name several alloys of which zinc is a constituent.

636. Name any defects occurring in brass, and state the causes to which these defects are due.

637. What are the properties which make brass so useful in the arts?

638.* Explain the terms "colouring" and "lacquering" as applied to brass articles.

639.* What metal replaces zinc in Japanese bronze, and with what results?

640. What is galvanized iron, and how is it prepared?

641. What is the nature of the alloy obtained from the bottoms of vats in which iron goods have been galvanized?

642.* Give the composition of a few useful bronzes used for coating brass.

643.* What are the composition and uses of lacquer employed for lacquering brass goods?

644.* In making brass a certain proportion of scrap brass is added to the charge. What purpose does this serve?

Lead.

645. Name the chief physical properties of lead.

646. What indication do specific gravity and the sound produced by percussion afford with respect to the relative purity of commercial varieties of lead?

647. What is meant by the term 'autogenous soldering'?

648.* When lead is exposed to moist air it becomes coated with a film. What is the nature of this film?

649. What are properties of protoxide of lead, and how is it produced?

650. What are the physical properties of 'massicot' and 'litharge,' and what is the difference between them?

651. What is the action of carbon, carbonic oxide, and of hydrogen on oxide of lead at a red heat?

652. What is the difference in composition between litharge and red lead, and how is the latter made on the large scale?

653. What impurities would you expect to find in litharge obtained from the cupellation furnace?

654. What is the influence of litharge on infusible oxides, such as ZnO, Fe_2O_3, and SnO_2, when strongly heated with them?

655.* Give examples to show that litharge acts as an oxidizing agent in some cases.

APPENDIX.

656. Mention the substances used as adulterations in red lead, and describe a simple experiment to show how they may be detected.

657. Describe by the aid of an equation the re-action which takes place when red lead and carbon are heated together at a red heat.

658.* When oxide of lead is heated in a clay crucible a fusible scoria is formed. What is its composition?

659. What change takes place when red lead is heated in a clay crucible (1) at a moderate temperature, (2) at a high temperature?

660.* What is silicate of lead, and how is it produced? How would you deal with it so as to extract the lead?

661.* What are the physical properties of silicate of lead, and for what purpose is it used in metallurgy?

662. What is the action of heat and air upon sulphide of lead?

663. What are the physical properties of sulphide of lead, and how may it be produced artificially?

664. Lead sulphide under certain circumstances is called galena, in other cases it is termed lead sulphide, while in others it is termed a regulus. Explain this.

665. Show by equations what occurs when sulphide of lead is roasted (1) at a low temperature, (2) at a strong red heat.

666. What re-action occurs when lead sulphide is fused with an alkaline carbonate? What is the difference when iron is also added?

667. Express by an equation the change which occurs when lead sulphide is strongly heated with iron.

668. In what metallurgical operation is lead sulphate formed, and what is its composition?

669.* What is the action of phosphorus and of arsenic respectively upon lead?

670. What is the nature of the substance known as white lead, and how is it formed?

671. What are the chief adulterations in white lead?

672. Mention the chief alloys of lead.

673. What is the composition of type metal, pewter, and Britannia metal, and for what purpose is each alloy employed?

674. What is the effect of zinc upon lead?

675. Suppose lead containing silver to be melted with zinc and allowed to cool slowly, what occurs? Give reasons for your answer.

676. Can silver and gold be alloyed with lead, and if so, is any use made of this property?

677. What important functions do lead and its compounds perform in the metallurgy of gold and silver?

678. Enumerate the chief ores of lead.

679. What is the principal British lead ore, and what are its physical properties?

Extraction of Silver from Lead.

680. Given a large quantity of lead containing about 200 ounces of silver per ton, how would you obtain the silver from it?

681. What are the principles of Pattinson's and Parkes' processes respectively?

682. How is silver extracted from galena?

683. How may silver be separated from lead on the large scale without the process of cupellation?

684. How would you ascertain the proportion of silver in lead without the use of chemicals?

685. In the blast furnace treatment of impure lead ores containing silver; 'slag,' 'regulus,' 'speise' and lead are sometimes obtained; where would you expect to find the silver?

686. Describe the method of desilverizing lead by means of zinc, and explain with the aid of sketches the appliances used.

687.* Give a sketch showing accurately in sectional elevation one of the pots with its furnace, used for concentrating lead by the Pattinson process.

688.* Discuss the influence of a process now used for desilverizing lead, in rendering the metal unsuitable for lining sulphuric acid chambers.

689. Describe the Cordurié method of desilverizing lead.

690.* Explain the method introduced by Rozan for desilverizing lead.

691.* Two systems are employed in the Pattinson process termed respectively the 'high' and 'low' systems. Explain these.

692. What is the nature of the products obtained from a Pattinson's pot, and for what purpose is each used?

693. What is the practical limit of concentration by the Pattinson process?

694. What is meant by cupellation, and what class of substances are treated by this method?

695. What is the nature of the furnace employed in England for the purposes of cupellation?

696. Give an outline of the process by which lead ores containing gold and silver may be smelted so as to extract the precious metals.

697. How may gold and silver be separated from copper in an alloy of these metals?

698. Wherein does the English differ from the German mode of cupellation?

699. Describe the German method of cupellation, and give a sketch in vertical section of the furnace employed.

700. In the cupellation of lead alloys, by what indications is the completion determined?

701.* What occurs (1) when the temperature is maintained too high, (2) when the temperature is kept too low, during the cupellation of argentiferous lead?

702.* In the desilverization of lead by means of zinc, some of the zinc remains dissolved in the lead. How does this influence the quality of the lead, and how may the zinc be eliminated from the lead?

Extraction of Lead from the Ore.

703. What is the principal ore of lead, and how is the metal usually extracted from it in the United Kingdom?

704. Classify the processes employed for the extraction of lead.

705. Describe the method of smelting galena in Flintshire, and give approximately the composition of the products.

706. Give a sketch in sectional elevation of the Flintshire furnace, and state the reasons for its particular mode of construction.

707. Give equations showing the chemical changes which occur when lead is extracted from galena by the Flintshire method.

708. At a certain period of the Flintshire process, lime is added; what purpose does the lime serve?

709.* Describe the method of lead smelting as carried out in Derbyshire.

710. How does the method of lead smelting in Brittany differ from the Flintshire process?

711. Describe the mode of working and nature of the furnace termed 'boliche,' employed in Spain for smelting galena.

712.* A modification of the Flintshire method is considered necessary for smelting certain Cornish lead ores; why is this?

713. Describe the furnaces used for smelting lead by the Cornish method.

714. What is the nature of the products obtained from the Cornish 'flowing furnace' in lead smelting?

715. Lime is added to the charge in calcining lead ores by the Cornish method; what is its action?

716.* When galena is smelted in a little furnace named the ore-hearth, state the chemical re-actions which take place.

717. Describe, with the aid of a sketch or sketches, the lead furnace known as the ore-hearth.

718.* How is rich galena smelted in England?

719.* Describe the Bleiberg or Carinthian process of lead smelting.

720. Describe the method of treating ores in the Freiberg district with a view to the extraction of lead.

721.* What circumstances would induce you to employ a blast furnace for treating lead ores, in preference to a reverberatory one?

722. Explain the part played by iron and by ferruginous 'cinder' or 'scoria' in the metallurgy of lead.

723. Describe the furnace employed for roasting lead ores at Pontgibaud, and state in what condition the material is removed from the furnace at the close of the process. What is the object of this roasting?

724. What is the condition of the ore in different parts of the furnace mentioned in the last question?

725. Show by the aid of a sketch in sectional elevation, the arrangement of a furnace with a water jacket, as used in the smelting of lead ores.

726. Describe the blast furnace employed for reducing roasted lead ore at Pontgibaud. Give equations showing the changes which occur.

727. What solution would you employ for depositing lead by electricity, and how would you proceed?

728.* Describe any method by which impure lead can be refined by electricity.

729. Lead produced from mixed ores is found to contain antimony, tin, copper and iron. Describe the process of purification.

730.* What methods may be adopted for treating the scum which rises to the surface of a bath of lead after the dezincification of the metal by the Cordurié process?

731. What is the composition of lead fume, and how may it be condensed and collected?

732. Describe the principles of the different methods which have been adopted for the condensation of lead fume.

733. State the properties and mode of producing slag lead.

734. How is hard lead softened? Explain the principles of the process.

735. Describe the method employed for reducing litharge on the large scale.

736.* How is lead piping made at the present time? Is such lead pure; if not, what are the metals alloyed with it?

737. Describe the method of treating lead slags so as to recover the lead, and state the nature of the products.

Tin.

738. What are the properties of tin which make it so useful in the arts ?

739. What influence does the atmosphere exert on tin (1) at the ordinary temperature, (2) at 300° C. ?

740. What is the melting point of tin ? How does this affect its application in the Arts ?

741. What simple method is used for ascertaining approximately the quality of commercial tin ?

742.* Describe the physical properties of stannic oxide, and state how it may be prepared from metallic tin.

743.* What is putty powder, and for what purposes is it used ?

744.* Describe the substance known as Mosaic gold, and state how it may be prepared.

745. In what alloys is tin used ?

746. What are gun metal, bell metal, bronze, and speculum metal ?

747.* Briefly describe the properties of the alloys mentioned in the last question.

748. What is the nature of the compound containing tin, known as 'hard-head' ? How is it produced ?

Extraction of Tin from its Ores.

749. Describe the substances from which tin is obtained.

750. Give a sketch of the furnace used in Cornwall for smelting tin ores, and describe the process of separating the metal.

751.* Describe the chemical re-actions which occur in smelting tin ores in a reverberatory furnace.

752. Describe the Cornish method of tin refining, and the theory of that process.

753. From what and how is grain tin produced in Cornwall ?

754.* A tin ore is found to contain wolfram. What alteration in the ordinary smelting process would its presence render necessary ?

755.* What chemical process has been adopted for the separation of wolfram from tin stone ?

756. How is the peculiar structure of grain tin produced ?

757. Explain the terms common tin, refined tin and block tin.

758. Sketch and describe the blast furnace used for smelting tin ores in Germany.

759. Describe the process of tinning iron plates, and state the purposes for which they are used.

NICKEL AND COBALT.

760. Describe the physical properties of nickel and cobalt respectively.

761. State the purposes to which nickel and cobalt are applied in the arts.

762. What is the action of carbon and of hydrogen on oxide of nickel ?

763. What are the nature and properties of the compound formed when nickel and sulphur are strongly heated together ?

764. What is the effect of arsenic on nickel at a red heat ?

765. What is nickel speise; how is it prepared, and what are its physical properties ?

APPENDIX. 231

766. What is the action of heat and air on nickel speise?

767.* Can you suggest a process for the treatment of nickeliferous iron pyrites on the large scale with a view to the production of nickel speise? If so, describe it, and state the chemical re-actions which take place.

768. What would take place if a compound containing both silicates of nickel and cobalt were fused with arsenical pyrites?

769.* How do you explain the production of nickel speise in the old process of manufacturing smalts?

770. If sulphides of nickel and cobalt be fused with an acid silicate of iron, what occurs?

771. If arsenides of nickel and cobalt be fused with an acid silicate of iron, what occurs?

772. If silicates of nickel and cobalt be fused with iron pyrites or sulphur, what occurs?

773. If silicates of nickel and cobalt be fused with arsenic or mispickel, what occurs?

774. Name the chief ores of nickel and cobalt.

775. What is German silver and nickel silver? Is there any difference between them?

776. Give the composition of three varieties of German silver, and state the purposes for which they are used.

777.* State the chief physical properties of German silver.

778.* Name any alloy in which cobalt is an important constituent.

779. What are the ingredients of cobalt blue?

780. Of what does the pigment termed 'smalt' consist, and how is it made?

781. What is the difference between 'zaffre' and smalt? How is the latter made on the large scale?

782. What foreign elements are generally present in commercial nickel ?

Methods of Extracting Nickel and Cobalt.

783. Describe the principles on which the extraction of nickel from its ores is based.

784.* How is nickel extracted from 'garnierite' or silicate of nickel ?

785.* In what manner would you treat a speise of copper, nickel, and cobalt, with a view to extract these metals ?

786.* What resemblances are there between the metallurgical treatment of (1) ores of iron and of nickel, and (2) ores of copper and of nickel ?

787. How is nickel separated from cobalt on the large scale, when those metals exist in the form of chlorides ?

788.* Describe a process for the treatment of arsenical ores of nickel and cobalt, and explain the re-actions which take place at each stage.

789. Explain how cobalt, nickel, and iron may be separated from each other in the smelting of ores containing them.

790.* Describe any method of extracting nickel from its ores by electrolysis.

791. How would you proceed to electro-plate an iron article with nickel, and what solution would you employ ? Give reasons for your answer.

Aluminium.

792. Describe the metal aluminium. What use is made of it in the arts ?

793. In what respects does aluminium resemble gold ?

794. What impurities occur in commercial aluminium?

795. How would you purify the crude aluminium as obtained from its ores?

796.* What is alumina, and how may it be artificially prepared? What are its chief properties?

797. How does aluminium occur in nature? What are the chief sources from which the metal is obtained?

798. What part does the metal sodium play in the metallurgy of aluminium?

799. Describe a recent method of preparing sodium at a cheap rate.

800. Describe an electrical method for obtaining aluminium.

801. What is aluminium bronze? Describe its properties.

802. Describe an electrical method of preparing aluminium bronze.

MERCURY.

803. In what metallurgical processes is mercury used?

804. What are the chief physical properties of mercury?

805. What is the effect of moist air on mercury (1) when perfectly pure, (2) when containing other metals?

806. What impurities are generally present in commercial mercury?

807.* Mercury has a high co-efficient of expansion for heat. Explain this, and state what use is made of this property?

808. Mention the chief practical applications of mercury.

809.* What is meant by 'water-gilding'? Explain the process.

810.* Explain the process of silvering mirrors by the aid of mercury.

811. What is vermilion, and how is it prepared on the large scale? How would you prepare a small quantity?

812. What substances are used to adulterate vermilion?

813. Give an equation to show the change which occurs when mercuric sulphide is heated with lime.

814. What is the effect of carbon and iron on vermilion at a red heat?

815. What is an amalgam? Mention some useful amalgams.

816.* How may mercuric oxide and mercuric chloride be prepared from metallic mercury.

817.* Describe the mode of preparing sodium amalgam, and state for what purposes it is used in metallurgy.

818. Name the principal ores of mercury, and describe the properties of the most important ore.

Methods of Extracting Mercury from its Ores.

819. State the principles of the different methods employed for extracting mercury.

820. How is mercury extracted from rich native cinnabar?

821. What method would you employ for extracting mercury from very poor ores?

822. Describe the construction and mode of working of the Alberti furnace for the treatment of mercury ores.

823. Describe the construction and the mode of working of the Hähner furnace.

824.* What is the theory of the reduction of cinnabar in the Almaden and Idrian furnaces?

825. How is crude mercury purified by the dry way, and what is the theory of the process?

826.* Describe any wet method of purifying mercury.

827.* Describe the method of extracting mercury by means of lime or iron. Give equations to express the chemical changes which occur in each case.

Antimony.

828. What are the chief physical properties of antimony?

829. In what metals in common use would you expect to find arsenic and antimony?

830. Describe the metal known in commerce as "regulus of antimony."

831. How is sulphide of antimony prepared, and what are its physical properties?

832.* What is the effect of passing steam over sulphide of antimony?

833. What is the action of carbon, iron, potassium cyanide, and sodium carbonate respectively, when fused with sulphide of antimony?

834. What is the chief ore of antimony? Describe its physical properties.

835. Give the composition, and describe the properties, of the alloys in which antimony is an important constituent.

836. What is the general effect of antimony on other metals in alloys containing it?

837. What is the appearance of antimony, and for what purposes is it used in the arts?

Methods of Extracting Antimony from its Ores.

838. How is antimony extracted from the native sulphide?

839. Describe a method of separating antimony from its ores by liquation.

840. Describe the English process of reducing antimony ores.

841. Describe the French method of extracting antimony, and sketch the furnace employed.

842. Describe the method of purifying the crude antimony of commerce.

843.* What is the nature of the substance termed "livers of antimony," and for what purpose is it used?

Arsenic.

844. What are the physical properties of arsenic?

845. What reasons are there for classifying arsenic amongst the metals?

846. How does arsenic occur in nature?

847. What is white arsenic and how is it prepared? How is it refined?

848.* From what and how is arsenical glass made? State its physical properties.

849. Describe the furnace employed for the production of arsenious acid from arsenical iron pyrites, and describe the re-actions which occur in the process.

850. How is metallic arsenic separated from its ores?

851. Name any useful alloys containing arsenic, and state the object of its introduction.

852. Describe the properties of realgar and orpiment.

BISMUTH.

853. What kind of a metal is bismuth, and for what purposes is it used in the Arts?

854. How is sulphide of bismuth prepared, and what are its physical properties?

855. Bismuth can be used instead of lead in cupellation. Explain how it acts and state what compound of bismuth is formed.

856.* Bismuth is sometimes obtained from cupels when the lead employed for cupellation has contained bismuth. How is the bismuth separated?

857. How does bismuth occur in nature?

858. Describe the method of extracting bismuth by liquation, and sketch the furnace employed.

859. For what purpose is bismuth chiefly used in alloys?

860. Describe some useful alloys containing bismuth.

861. What is the composition of Wood's alloy? What is its melting point?

862.* Describe the method employed for the reduction of bismuth from an ore containing disseminated sulphide of bismuth.

INDEX.

	Page		Page
A		Asphalte,	34
		Atacamite,	110
Acid substance,	20	Augustin's process for silver,	93
After-blow,	84	Azurite,	110
Aich's metal,	128		
Air gas,	25		
Alloys,	12	**B**	
Aluminium, alloys of,	160		
,, electrolysis of,	160	Barff's process,	50
,, extraction of,	158	Barker's process for gold,	102
,, ores of,	158	Barrel furnaces,	21
,, properties of,	158	Basic substance,	20
,, bronze,	160	,, process,	83
Amalgamation,	6, 88	Bath metal,	128
,, electrolytic method,	102	Bauxite,	20, 56
		Bears,	65
,, German method,	90	Bell metal,	148
,, Jordan's ,,	101	Bessemer acid process,	80
,, Mexican ,,	88	,, basic process,	83
Amalgams,	162	Bicheroux furnace,	23
Anode,	97	Bismuth, alloys of,	172
Anthracite,	33, 34	,, properties of,	171
,, ashes of,	33	,, glance,	171
Antimony, alloys of,	168	,, ochre,	171
,, liquation of,	169	,, ores,	171
,, ores of,	168	,, ,, liquation of,	172
,, ores, reduction of,	169	Blackband,	45
,, properties of,	167	Black Jack,	124
Appolt's coke oven,	40	Blast furnace gases,	64
Arsenic, alloys of,	171	Blende,	124
,, extraction of,	171	Blister steel,	76
,, ores of,	171	Blowholes,	80
,, oxides of,	170	Blue billy,	46
,, properties of,	170	Boetius' furnace,	23
,, sulphides of,	170	Boghead,	34
,, white,	170	Brass,	128

INDEX.

	Page		Page
Brass, properties of,	128	Coking in kilns,	38
Britannia metal,	168	,, ovens,	38
British bronze coinage,	147	,, piles,	37
,, gold ,,	99	Conductivity,	10
,, silver ,,	87	Coppée's coke oven,	40
Bronze,	147	Copper, compounds of,	109
Brown hæmatite,	45	,, extraction of,	110
Bull dog,	47	,, extraction of, by	
,, ,, slag,	47	electricity,	121
		,, ores of,	110
		,, properties of,	108
C		,, glance,	110
		,, plating,	123
Calamine,	124	,, pyrites,	110
Calcination,	4	Cowles' method for aluminium,	160
Calorific intensity,	29	Cowper's stove,	62
,, power,	27	Cox's coke oven,	39
Cannel coal,	34	Crucibles,	18
Carat,	98	Cryolite,	159
Case hardening,	76	Crystallization,	14
Cassel's process for gold,	102	Cup and Cone,	60
Castilian furnace,	141	Cupel,	6
Cast iron, grey,	48	Cupellation,	6
,, ,, mottled,	48	Cupola furnace,	60
,, ,, white,	48	Cuprite,	110
,, ,, heating stove,	61		
Cast steel,	77	**D**	
Catalan process,	54		
Cathode,	97	Delta metal,	128
Cementation,	7	Desulphurization of coke,	43
,, process,	75	Dinas clay,	20
Cerussite,	133	Distillation,	5
Charcoal,	35	Dolomite,	20
Chimney,	22	Dome furnace,	22
Chromium in iron,	49	Dry process,	13
Chrysocolla,	110	Ductility,	10
Cinnabar,	163		
Clapp-Griffith's converter,	81	**E**	
Claudet's process for silver,	119		
Clay ironstone,	45	Elasticity,	10
Cleveland ironstone,	45	Electric calamine,	124
Coal,	31, 32	Electro-amalgamation,	102
,, ashes of,	33	Electro-chemical process,	14
Cobalt, alloys of,	156	Electro-deposition,	96
,, ores of,	156	Electro-gilding,	103
,, properties of,	156	Electro-plating,	97
,, bloom,	156	European-amalgamation	
,, glance,	156	process,	90
,, speise,	156	Examination questions,	173
Coke,	37	Expansion,	9

INDEX.

F

	Page
Fahl ore,	110
Ferro-manganese,	49
Finery,	71
,, ball,	71
,, slag,	72
Fire-bricks,	17
Fire-clay,	20
Flue,	22
Fluor-spar,	17
Flux,	16
Forge,	20
Fracture, varieties of,	13
Franklinite,	44
Freiberg process,	91, 92
Fuel,	27
Furnaces,	20
,, classification of,	21
Fusible alloy,	172

G

	Page
Galena,	133
Gallery furnace,	22
Galvanized iron,	48
Ganister,	20, 81
Gangue,	3
Garnierite,	153
Gas furnaces,	23, 70
Gases of coke ovens,	42
Gedges' metal,	128
German silver,	152
Gerstenhöfer calciner,	115
Gilding,	103
,, metal,	128
Gjers' calciner,	52
Gold, alloys of	98
,, coloured,	99
,, ores of,	100
,, parting of,	103
,, pure,	99
,, ores, amalgamation of,	100
,, ,, electro-amalgamation of,	102
,, ,, electro-deposition of,	102
,, ,, Jordan's process,	101
Göthite,	45

	Page
Graham,	6
Grey copper ore,	110
,, pig iron,	48
Gun metal,	148

H

	Page
Hammer, helve,	68
,, steam,	69
,, tilt,	68
Hardness,	11
Hearth,	20
Heating by conduction,	23
,, ,, regeneration,	24
Henderson's process for copper,	120
Hollow fire,	71
Hot air,	23
Hungarian mill,	100

I

	Page
Ilmenite,	45
Indian steel,	77
Iron,	44
,, alloys of,	48
,, chemistry of,	46
,, compounds of,	46
,, extraction of,	52
,, native,	44
,, ores of,	44
,, properties of,	49
,, pure,	46
,, blast furnace,	60
,, ,, ,, charges of,	57, 59
,, ,, ,, chemical action in,	58
,, ,, ,, slag produced in,	59
,, ores, calcination of,	51
,, ,, preparation of,	51
,, ,, roasting of,	51
,, pyrites,	45

J

	Page
Jet,	34
Jordan's process for gold ore,	101

INDEX.

	Page
K	
Kaolin,	20
Kernel roasting,	119
Kupfer nickel,	153
L	
Lead, alloys of,	132
,, cupellation of,	135
, electrolytic method,	143
,, extraction of, in Brittany,	138
,, ,, Cornwall,	139
,, ,, Flintshire,	137
,, ,, Ore hearth,	140
,, ,, Spain,	139
,, fume,	143
,, ores of,	133
,, oxides of,	131
,, properties of,	130
,, separation of silver from,	133
,, sulphide of,	131
,, smelting at Freiberg,	140
,, ,, Pontgibaud,	141
,, ,, in the Hartz,	142
,, ,, Sweden,	140
,, softening,	144
Lignite,	32
Litharge,	130
,, reduction of,	145
Liquation,	5
,, process,	87
M	
Magnetite,	44
Malachite,	110
Malleability,	10
Malleable cast iron,	79
Manganese in iron,	45
Martin process,	77
Massicot,	131
Mercury, amalgams of,	162
,, compounds of,	162
,, extraction of,	164
,, ores of,	163

	Page
Mercury, properties of,	162
,, purification of,	166
Metallurgy,	1
Metals, melting points of,	8
,, properties of,	7
,, useful,	2
Mexican amalgamation process,	88
Mill,	71
,, furnace,	69, 70
Mimetesite,	133
Mosaic gold,	128
Muffle furnaces,	21
Mushet steel,	77
N	
Neutral substance,	20
Nickel, alloys of,	152
,, extraction of,	153
,, ores of,	153
,, properties of,	152
,, glance,	153
,, plating,	154
,, pyrites,	153
,, silver,	152
,, speise,	152
Nielson,	61
O	
Occlusion,	6
Odour,	10
Open fire,	71
Open hearth,	24
Open hearth steel,	77
Ore, definition of,	3
Orpiment,	170
P	
Parting of gold from silver,	103
Parkes' process,	135
Patio-amalgamation process,	88
Pattinson's process,	133
Peat,	31
,, ashes of,	31

INDEX.

	Page
Peat, charcoal,	35
Pernot's furnace,	79
Petroleum,	34
Pewter,	132
Pilz' furnace,	91
Pinchbeck,	128
Plating vat,	97
Platinum,	105
,, electro-deposition of,	106
,, extraction of,	105
Platinating,	107
Platinizing,	106
Poling of copper,	113
,, tin,	150
Ponsard's furnace for re-heating,	70
,, ,, for steel,	79
Potassium cyanide,	97, 107
Prepared fuel,	34
Prince's metal,	128
Printer's blue,	157
Puddled balls, treatment of,	68
,, steel,	74
Puddling, dry,	66
,, wet,	67
,, furnace,	68
,, ,, re-actions in,	67
Pyrometers,	30
Pyromorphite,	133

R

	Page
Rachette furnace,	142
Realgar,	171
Recuperator,	23, 70
Red hæmatite,	44
Red lead,	131
,, manufacture of,	145
Reducing agent,	4
Reduction,	3
Refinery,	66
Refining pig iron,	65
Refractory materials,	17
Regenerative furnace,	24
Regulus,	3
Re-heating,	69
,, furnace,	69

	Page
Reverberatory furnace,	21
,, ,, atmosphere of,	21, 22
Roasting,	5
Rotatory furnace,	56

S

	Page
Scaffolds,	65
Scoria,	6
Scorification,	6
Scouring slag,	59
Shot metal,	132
Sick mercury,	102
Sickening,	102
Siderite,	45
Siemens' gas furnace,	24
,, gas producer,	25
,, regenerative furnace,	24
,, rotatory furnace,	56
,, steel process,	77
Siemens-Martin process,	78
Silicate of lead,	6
,, lime,	6
Silicates, classification of,	16
Silver,	85
,, alloys of,	87
,, compounds of,	86
,, extraction of,	87
,, ores of,	87
Simon-Carvè's coke oven,	41
Slag,	15
Slags, properties of,	16
Slips,	65
Smalt,	156
Smaltine,	156
Sodium amalgam,	163
Soft solder,	132
South Wales finery,	71
Spathic iron ore,	45
Specific gravity,	11
,, heat,	11
Speculum metal,	168
Specular iron ore,	44
Speise,	3
Spiegel-eisen,	49
Stagg's condenser,	144
Steel,	73

INDEX.

	Page
Steel, casting of,	80
,, impurities in,	73
,, production of,	74
Stereotype metal,	168
Sterro metal,	128
Stibnite,	168
Sublimation,	5
Sulphides,	17
Sulphur in coal,	33
Swedish-Lancashire finery,	72
Swedish copper furnaces,	117

T

	Page
Tap cinder,	68
Taste in metals,	10
Temperature,	8
Tenacity,	9
Terne plate,	147
Tharsis method for copper,	120
Thomas-Gilchrist process,	83
Tin, alloys of,	147
,, block,	150
,, common,	150
,, deposition of,	151
,, extraction of,	149
,, grain,	150
,, ores of,	149
,, properties of,	147
,, tossing of,	150
Tinstone,	149
Tombec,	128
Toughness,	9
Trial bar,	75
Trompe,	54
Trunnions,	81
Type metal,	168

V

	Page
Vermilion,	162
Volatile metals,	8

W

	Page
Waste gases,	64
,, ,, Langen's apparatus,	65
,, ,, cup and cone,	61
Water gas,	25
Welding,	10, 50
Wet process,	13
White lead,	146
White pig iron,	48
Whitwell's stove,	63
Wood,	31
,, ashes of,	31
,, charcoal,	35
Wood's alloy,	172
Wootz steel,	77

Z

	Page
Zaffre,	157
Ziervogel's process for silver,	95
Zinc, alloys of,	128
,, electrolytic extraction of,	127
,, extraction of, in Belgium,	126
,, ,, ,, England,	126
,, ,, ,, Silesia,	127
,, ores of,	124
,, properties of,	123
,, blende,	124
Zincite,	124

www.ingramcontent.com/pod-product-compliance
Lightning Source LLC
Chambersburg PA
CBHW021403230426
43666CB00006B/626